Anonymous

Proceedings of the Victoria Institute of Trinidad

Anonymous

Proceedings of the Victoria Institute of Trinidad

ISBN/EAN: 9783337321000

Printed in Europe, USA, Canada, Australia, Japan

Cover: Foto ©berggeist007 / pixelio.de

More available books at **www.hansebooks.com**

VICTORIA INSTITUTE,

TRINIDAD.

REPORT FOR THE YEAR 1894.

THE Board of Management of the Victoria Institute have to report that during the past year lectures were given as follows :—

1894.		Subject.	Lecturer.
1 February	...	Music	W. S. Doorly
21 February	...	Botany	J. H. Hart
5 April	...	Music	W. S. Doorly
23 August	...	Zoology	R. J. L. Guppy
15 November	...	Botany	J. H. Hart
13 December	...	Zoology	R. J. L. Guppy

On no occasion was the attendance at the lectures given by the Institute satisfactory, and on many occasions the lecturers had the trouble of preparing their lectures for no purpose.

The following are the dates of the scientific meetings with a note of the business done at each meeting.

1 February—Microscopical Evening.

15 April—On Cutting Timber in Trinidad.   ...R. J. L. GUPPY

3 May—Discussion on Cane Disease.

14 June—On the Helicinidæ.   ...   ...R. J. L. GUPPY

9 August—Notice of a Paper on the Mammals of
Trinidad by J. A. Allen and F. M.
Chapman ; A Notice of a Paper on
the Birds of Trinidad by F. M.
Chapman; and a Notice of a Paper
on the Orthoptera of St. Vincent
and Grenada; also Bibliographical
references to recent scientific
papers relative to Trinidad and the
neighbouring islands.

29 November—Physical conditions and Fauna of
Gulf of Paria...   ...   ...R. J. L. GUPPY.

27 December—Molluska of Gulf of Paria   ...R. J. L. GUPPY.

As fully explained in last Report the Meetings and Lectures
are practically free to all who desire to attend them. Persons
who are not members or associates can always find some one who
is a member to give them the required ticket or introduction.

The number of Members and Associates on the Books of the
Institute at 31st December last was as follows :—

Life Member...   1

Non-paying Members 5

Paying Members who have paid up..52

„   „   who have not paid up.. 9

Associates 4

Grand Total   .. 71

It is much to be regretted that the class of Associates has not developed itself more extensively. This is the class for whom the lectures are chiefly intended and for whose benefit technical instruction is proposed to be given in the Institute. It should consist largely of the young men in offices, stores, counting-houses, training-schools and colleges, and engaged in mechanical and industrial pursuits. All the real advantages of the Institute are open to this class, the subscription to which is only Ten Shillings a year.

The Institute has lost by death during the year two Members who filled prominent places in the community, namely :

BEAVEN NEAVE RAKE, M.D., &c.

The Hon'ble ROBERT GUPPY, M.A.

Obituary Notices of these gentlemen are appended to this Report.

The number of persons who entered their names in the Visitors' Book of the Museum during the past year was as follows :—

January	...	... 207	July	... 675
February	...	78	August ...	... 809
March	...	... 65	September	... 633
April		... 137	October ...	... 474
May		... 874	November	... 605
June	...	... 544	December	... 521

Making a total of 5,672 Visitors (besides small children and others) during the year.

One of the functions inherited by this Institute from its predecessor the Scientific Association is that of placing on record as far as possible all that is known of the fauna and flora of the island and its neighbourhood. As part of the Journal of Proceedings of the Institute for the current year notices of papers on scientific subjects which have been before meetings of the

Institute will be published. This function is among the most useful of those generally performed by local scientific bodies. Such lists as are appended to these notices serve as indexes to the literature of the subjects and are also convenient as check lists enabling one at a glance to take in the state of our knowledge, and to see as far as that knowledge goes what the composition of the fauna (or whatever branch may be treated of) is, and to compare the species represented or other subjects treated of with those of other places or collections.

An application was made during the year to His Excellency the Patron for the grant to the Institute of copies of Government Publications. His Excellency was good enough to accede to the application and in consequence the Institute has received from the Government Printing Office the following volumes:—

Trinidad Official Gazette... ...	... 34	Volumes.
Council Papers and Minutes ...	... 23	,,
Yearly Blue Books of the Colony	... 10	,,
In all ...	... 66	,,

These form a notable addition to the Library of the Institute and will be exceedingly useful for reference.

The Institute has received donations of Books, &c. from various persons and bodies as detailed in the Proceedings.

A beginning has been made in binding the books left to the Institute by the Scientific Association.

Correspondence has been carried on during the year with the Imperial Institute of Great Britain, the Colonies and India with a view to improving our relations with the Institute ; but it cannot be said that any definite course of action has yet been decided on.

In April last it was discovered that the Clerk and Assistant Curator, Gibbs, had committed serious frauds by embezzling moneys paid as subscriptions to the Institute. The police were communicated with on the subject, but as Gibbs absconded at once on discovery, no further proceedings could be taken.

A proposal for an exhibition of Sketches in oils and water colors and in black and white and of Photographs was lately brought before the Board of Management. This was approved and the exhibition was held on the 17th, 18th and 19th January 1895. Although the exhibition was held therefore during the current and not during the past year, yet as it was the outcome of arrangements made during 1894 it may well be noticed in connection with the report for that year.

About 250 sketches and paintings were sent in besides photographs of which Mr. Sellier exhibited a fine series. As the Hanging Committee were able with some difficulty to find space for the whole of the exhibits, none were rejected, although a few were of doubtful merit. Most of the articles sent in showed taste or ability in a greater or less degree and a few were decidedly of a high order. It is not intended to go into any detail on this subject, but it may be mentioned that Lady Broome was good enough to lend to the Exhibition a sketch in colours by the late Captain Buckle and several other pictures including two miniatures by Zoffany. The exhibition was primarily and chiefly intended for the works of local artists and these formed the great majority of the exhibits ; but other works were not excluded. His Excellency the Governor and Lady Broome visited the exhibition on the afternoon of the first day. It was open on the afternoon of the 17th, 18th and 19th and also on the evening of the last day. It is estimated that about 800 persons visited the exhibition, and the receipts were £14 13 11 of which £10 7 2 was for admissions at the rate of sixpence for each adult and threepence for each child. Members of the Institute were admitted free and every exhibitor also received a free ticket. The expenses were £8 18 5 and the net proceeds amounted to £5 15 6 which sum is carried to the credit of the funds of the Institute for the year 1895.

The Governor and Legislative Council have been good enough upon the representation of the Board to grant an increased subsidy to the Institute. A vote of £250 was taken for the Institute for this year. We hope that this vote will enable us to make some progress in carrying out the objects of the Institute.

The Accounts and Balance Sheet furnished by the Treasurer show that the receipts of the Institute for the year 1894 amounted to $619.78 and the payments to $534.95. The receipts from subscriptions during the year showed a slight falling off, more than accounted for by the defalcations of the late clerk. Reimbursements were considerably less in 1894 than in 1893 owing to the fact that while in 1893 $150 was received from the Chicago Commission for rent, the amount received under the same head in 1894 from the Agricultural Society came to $60 only. Against a decrease of receipts the Board has to set an expenditure decreased in larger proportion, so that while the balance to credit to the Institute at the beginning of the year was $131.53 at the end of the year it was $216.36 an increase of $84.83. With so small an income as the Institute had during 1894 nothing could be done towards the reduction of the debt due to Government, but the increased grant voted by the Legislative Council alluded to above will put the Board in a position to begin payment of the instalments of the debt, and the Board trusts that the vote will be continued by the Council when the whole debt will be paid off in the course of five or six years.

R. J. LECHMERE GUPPY.

Victoria Museum,
    Port-of-Spain,
        30th January, 1895.

Statement of the Receipts and Expenditure of the Victoria Institute for the Year 1894.

RECEIPTS.		$	c.
Government Grant	...	240	00
Subscription of Members, &c.	...	240	90
Reimbursements for Meetings, &c,	...	138	88
		$619	78

JOHN T. GOLDNEY, *President.*

PAYMENTS.		$	c.
Salaries and Wages	...	243	40
Repairs—Furniture and Fittings	...	52	30
Petty Expenses and Lighting	...	93	25
Advertising, Printing, Postage and Stationery	...	146	00
		$534	95

R. J. LECHMERE GUPPY, *Treasurer.*

Victoria Institute—Balance Sheet 31st December, 1894.

Balance 1st January, 1894				
Cash in Bank	...$137 58			
Less due Treasurer	... 6 05		$131	53
Receipts as per Statement above			619	78
			$751	31

JOHN T. GOLDNEY, *President.*

Expenditure as per Statement above		...	$534	95
Balance 31st December, 1894.				
Cash in Bank	... $237 08			
Less due Treasurer	... 20 72		216	36
			$751	31

R. J. LECHMERE GUPPY, *Treasurer.*

I find this Cash Account correct according to the Vouchers presented.

G. CREAGH-CREAGH, *Auditor.*

I have compared the above Account with the Vouchers and find it correct.

C. W. LANGFORD, *Auditor.*

OBITUARY NOTICE OF BEAVEN NEAVE RAKE.

THE death of BEAVEN NEAVE RAKE, M.D., took place at Le Chalet, Maraval, Trinidad, on 24th August last. He was Vice-President of the Victoria Institute and Government Member of the Board of Management. Beaven Rake was born 28th April, 1853, consequently, he was only in his 37th year when he died. The following notice of his career is taken from the *British Medical Journal.*—

Beaven Rake's record has been a brilliant one, both as a student and since his qualification in medicine. Entering at Guy's Hospital in 1874 he soon made his mark at that school. He took several prizes, one at the Physical Society for his essay on the Localisation of the Functions of the Brain, as well as the Joseph Horace Prize (1877), and the Gurney Horace Prize (1879). He early matriculated at the University of London, and passed the Preliminary Scientific in the first division and with honours in 1876. In the following year he passed the first M.B.. also with honours, and subsequently took the M.B. degree with first class honours in medicine and honours in obstetric and forensic medicine. In 1882 he became M.D. Lond., with marks qualifying for the gold medal. He took the M.R.C.S. in 1879, and L.R.C.P. in 1880. He held several appointments at Guy's, house surgeon in 1882, house physician in 1883, and also for some time the post of resident obstetrical assistant.

After leaving the hospital he spent a considerable time abroad studying the special departments in Vienna, Berlin, and Paris. On his return to London he was appointed medical and surgical registrar at the Victoria Hospital for Children, and in 1887 accepted a nomination for a colonial surgeoncy in Trinidad. He was at once placed in charge of the Leper Hospital at Coco-rite, and entered upon his duties and on the scientific study of the subject of leprosy with that industry, energy, and ability which had characterised all his previous work. His first com-munication on the subject was to the Pathological Society of London in 1885 on Tuberculous Leprosy of the Tongue and Larynx. During the following year he made seven communi-cations to that Society, and since then many instructive specimens and valuable observations on the bacilli, inoculation experiments, etc., have been recorded by him in the Pathological *Transactions*.

It would take up too much space here to enumerate all the papers which he contributed to the various congresses, societies, and journals in this and other countries. Suffice it to say that in every one of them there is something original and thoughtful, and, like his able *Annual Reports of the Leper Hospital*, they are all full of careful work. We may say that several of his most important communications were presented at the annual meet-ings of the British Medical Association, and have been published in the *British Medical Journal*.

When, in 1890, the Committee of the National Leprosy Fund determined to send three commissioners to investigate the question of leprosy in India, the Royal College of Physicians was requested to nominate one of the Commissioners. At the suggestion of the writer of this article the late Sir Andrew Clark offered by telegram the post to Beaven Rake, who, with the permission of the Colonial authorities, promptly accepted the position. It is not too much to say that Dr. Rake was the strongest and most experienced member of that Commission, and that his name has added much to the weight and value of its report.

OBITUARY NOTICE OF THE HON. ROBERT GUPPY.

THE Hon. ROBERT GUPPY, M.A., Member of the Legislative Council, died at his residence, Piedmont, San Fernando, on the 12th November, 1894.

Mr. Robert Guppy was the third son of Samuel Guppy, Esquire, of the City of Bristol, and of Arno's Court, Somerset-shire, England.

Mr. Guppy was born in Bristol on 22nd January, 1808. He was educated at private schools in Leicester and London, and also, for six months, in the College of Henry IV in Paris. He then entered at Pembroke College, Oxford, where he graduated B.A., in 1829, and M.A., in 1830, and having duly kept the terms at the Middle Temple was called to the Bar in November, 1831, and went the Oxford Circuit, and at intervals travelled much in France, Switzerland and Italy.

In 1834, he married the only daughter of Richard Parkinson, Esquire, of Kinnersley Castle, Herefordshire, grand-daughter of Admiral Lechmere, and niece of the second Lord de Sausmarez.

Mr. Guppy practised at the Bar in England for some time after his marriage. Then at the request of Philip Protheroe, Esquire, (a connection by marriage) he came to Trinidad in the year 1839, to endeavour to retrieve the properties, then extensive, of Messrs. Protheroe and Son. Having discharged his mission he remained in the island, and was for a time connected with the sugar interest.

Mr. Guppy enjoyed the friendship and esteem of Sir Henry McLeod by whom he was appointed a Justice of the Peace and Road Commissioner in the year 1840 ; and in the year 1844 he was sent by him under the instructions of the Colonial Minister on a mission to Sierra Leone in the barque *Senator* (then employed in bringing African Immigrants to Trinidad) to inquire into the prospects of obtaining a sufficient supply of labourers from that settlement. And his report and advice had a great share in promoting the introduction of coolies from India : the first importation of whom took place by the ship *Fatel Rosak* in 1846 under the administration of Sir Henry McLeod. Upon his return from Africa Mr. Guppy was appointed by Sir Henry McLeod as acting Stipendiary Justice for the Naparimas and Savana Grande, to which the Couva district was added upon the death of Mr. David. The duties of that immense district, throughout which there was not then a load of gravel upon any road and no bridge over any river except the Couva river, were performed without a single day's absence and to the complete satisfaction of the authorities.

This acting appointment continued for fifteen months, at the expiration of which Mr. Guppy was appointed first Town Clerk of San Fernando under the Ordinance just then passed under the auspices of Sir Henry McLeod for constituting a Municipal Corporation in that town.

Soon after the arrival of Sir Charles Elliott and, upon the remodelling of the Territorial Ordinance (first enacted under Lord Harris), Mr. Guppy was appointed Warden of North Naparima Ward Union. And having had ample experience of the necessity of better means of communication (whilst the attempts made in Lord Harris's time to obtain a general system of Railways had utterly broken down) with the cordial assistance of Mr. William Eccles, Mr. Richard Darling, Dr. Philip and other eminent local proprietors of that

day, he promoted and succeeded in carrying out the Cipero
Tramroad (a railway under another name) by. means of which
Princestown has become the most prosperous internal town
in the island. Mr. Guppy was Secretary and General Super-
intendent and Manager of this concern until its completion
in 1867.

In 1862 Mr. Guppy had resigned his post as Warden
and on the completion of his engagement with the Commissioners of the Cipero Tramroad he resumed his practise as a
Barrister.

Practically, the first introduction of Railways into Trinidad
was due to Mr. Guppy. No less was this the case with the
electric telegraph, the first introduction of which into the
island was by his hands in 1868. Mr. Guppy's proposal to
connect the towns of San Fernando and Port-of-Spain by
telegraph was greeted by our Legislators in much the same
manner as George Stephenson's proposal to run trains drawn
by steam engines was in the House of Commons. We need
not add that it is many years since the proposed connection
became an accomplished fact.

Mr. Guppy was for many years Mayor of San Fernando.
He took an active part in all measures for the well being of
the Colony and he devoted a large portion of his life to the
service of his fellow townsmen and his fellow countrymen.

———————

The above notice compiled by a member of Mr. Guppy's family
has appeared in most of the local papers.

Proceedings of the Victoria Institute.

THURSDAY, 15TH APRIL, 1894.

Sylvester Devenish, M.A., Vice-President, in the Chair.

A paper on Cutting Timber in Trinidad by R. J. Lechmere Guppy, was read. The following is an abstract thereof.

ON CUTTING TIMBER IN TRINIDAD.

BY R. J. LECHMERE GUPPY.

(Abstract.)

The author first called attention to the different characters of Forest and Timber Trees in different countries and climates. Through a great part of Australia the woods were open while the timber was hardwood and involved somewhat different treatment to the timber of other temperate climates. In New Zealand the timber was very fine and belonged chiefly to the Coniferous division of Plants. Here immense forests had been destroyed by fire leaving only in places islands as it were of the grand trees known as kahikatea and other gregarious pines. The totara, a wood much like cedar, was not quite so gregarious. The author gave some particulars as to the modes in use of cutting down and cutting up these trees. He referred to the use of the axe, the saw, (cross-cut and pit) to the setting out of the lengths of logs, the procedure for getting the logs on to the pit, the breaking-down and flitching of the log when on the pit, &c., &c. He then contrasted these methods with those in use so far as his own observations went in Trinidad. He referred to the destruction of timber, particularly balata, on our northern hills, and suggested for consideration whether something in the way of portable saw-mills driven by water-power might not be used for working up the balata and other woods. He then referred to the mode of dealing with the West Indian cedar and quoted the following note published by him some years ago in the Trinidad Register and Almanack.

The mode of cutting timber in Trinidad is generally a wasteful one, especially as regards the valuable wood called cedar (*Cedrela odorata*). The rudeness of the structure which usually serves for a sawpit is hardly perhaps to be avoided, as it is seldom that more than a single tree can be had within a radius permitting the use of one pit. Little skill is shewn in the setting out of the lengths of the logs, and severing with the axe instead of cross-cutting with the saw is often employed. After this is done comes a greater waste. Instead of getting the log on to the pit, and there breaking-down or flitching it, as might be ascertained to be best, it is roughly squared with the axe before being put on the pit, whereby something like half the wood is with great labour reduced to chips only fit for fuel.

A discussion followed the reading of the paper in which the Chairman and others took part and in the course of which different opinions were expressed on the subject of the paper.

THURSDAY, 3RD MAY, 1894.

His Hon. Sir John Goldney, President, in the Chair.

The following Donations were exhibited ;—

A specimen of *Anodon Leotaudi* presented by C. W. Meaden, Esq.,

A pouched Rat (*Heteromys anomalus*) in spirit—presented by J. H. Hart, Esq., F.L.S.

Specimens of Cacao of Nicaragua.
  Theobroma bicolor   „

Mangrove Cutch ⎫ Manufactured at the Botanic
               ⎬ Gardens—Presented by J. H.
Bermuda Arrowroot ⎭ Hart, Esq., F.L.S.

Specimen of Sand from the Island of Ascension formed of rounded fragments of shell—Presented by Mrs. H. D. Huggins.

Specimen of Gordius (dry)—Presented by Lechmere Guppy, Esq, jnr.

Indian Pestle—Presented by P. Darcueil, Esq.

––––––––

The subject of Cane Disease was discussed by the members with special reference to the description of *Trichosphæria sacchari* by Mr. Massée. The opinion was generally expressed that the most effective means of keeping in check this and other pests would be found in burning all cane refuse from affected fields.

On the subject of Cane Disease reference may be made to a paper by Ernest Francis, in Proc. Scient. Assoc. Trinidad, 1879, page 183.

TnURSDAY, 14TH JUNE, 1894.

IIis IIonour Sir John Goldney, President, in the Chair.

The following Donations were announced :—

Twelve Volumes of Guide Books to the collections of the British Museum (Natural History).   *The Trustees.*

Massée on Trichosphœria Sacchari—*The Colonial Secretary.*

Boletin del Instituto geografico Argentino, tom XIV., euad. 9—12.   *The Institute.*

Elisha Mitchell Scientific Society—Proceedings, 8 parts *The Society.*

Botanic Gardens—Reports 1887—93; and Bulletin, 12 parts.   *J. II. Hart. F.L.S.*

The following paper was read :

ON A LANDSHELL OF THE GENUS HELICINA FROM GRENADA AND ON THE CLASSIFICATION OF THE HELICINIDÆ.

BY. R. J. LECHMERE GUPPY.

By the kind intervention of Mr. F. W. Urich I have received from C. W. Smith of Grenada, specimens of a *Helicina* found in that Island and named *II. Keatei* by that great authority on landshells the late Louis Pfeilfer of Cassel. Though acquainted with the shell, I had not previously possessed a specimen nor is there one in the collection of the discoverer the late Governor Keate, after whom it is named. This shell belongs to a small group which may be typified by *II. caracolla* of Brazil but its nearest relation is *II. occidentalis* of St. Vincent.

The HELICINIDÆ to which our shell belongs is a family of operculate landshells consisting of the genus Helicina (taken in a very wide sense) and of the two allied *genera Preserpina* and *Ceres.* Viewed in this way the family is an exceedingly natural and well defined group without near allies. Among land shells its nearest relations would be among the *Cyclostomidæ*. Among marine molluska affinities are indicated both as to shell characters and dentition with *Neritina* and through that genus probably with *Trochus, Haliotis* and *Fissurella*.

It would, I think, be a great convenience for zoological study and classification if every family or genus or other group of animals were treated somewhat in the way that the family Helicidæ is dealt with in the second edition of Albers' work " Die Heliceen." Unfortunately some writers or naturalists seem determined to spoil every attempt to deal philosophically with classification and nomenclature. Albers' work is doubtless full of errors and will bear much amendment. But the principle on which it is based, that of giving as wide an extension as nature may admit to generic names and dividing each genus into sections, is a most useful one. The names given to these sections are not intended to be generic names nor ought they to come strictly under the laws applicable to generic names. But some authors have unfortunately quite misunderstood this point and have adopted Albers' names as names of genera, whereon an immense amount of ink and labor has been wasted in applying to such names the rules of priority, &c.

I exhibit to you a small but valuable collection of Helicinidæ which I have attempted to arrange in the manner indicated and I append a list of the species, showing the proposed genera, subgenera and sections into which the species may be grouped.

Trinidad possesses four species of the genus *Helicina* all very distinct from one another and belonging each to a different section of the genus.

GENUS HELICINA LAMARCK

Subgenus LUCIDELLA Gray

Helicina aureola Fér.   Jamaica

Subgenus STOASTOMA Adams

Helicina pisum Adams   Jamaica

Subgenus Trochatella Swainson

Helicina tankervillei Gray	Jamaica
pulchella Gray	,,
josephinœ Adams	,,
sloanei Orb.	Cuba
petitiana Orb.	,,
candeana Orb.	Venezuela

(Fitzia)

Helicina regina Morelet	Cuba

Subgenus Schasicheila Shuttleworth

Helicina pannucea Morelet	Guatemala
alata Menke	Mexico
bahamensis Pfeiff.	Bahamas

Subgenus Helicina s.s.

(Perenna)

Helicina lirata Pfeiff.	Mexico
lineata Adams	Jamaica
semistriata Sow.	Panama
lamellosa Guppy	Venezuela, Trinidad

(Poenia)

Helicina plicatula Pfeiff.	Dominica, S. Lucia &c.
rugosa Pf.	Cuba
ignicoma Guppy	Trinidad
conoidea Pf.	Barbados
paivana Pf.	Haiti

(Urichia)

Helicina adamsiana Pfeiff.	Jamaica
coronula Shuttl.	,,

(Ampullina)

Helicina moquiniana Pf.	Fiji
amœna Pf.	Guatemala
concentrica Pf.	Venezuela

(DIAPHANA)

Helicina subfusca Menke S. Thomas, W.I.

(KREBSIA)

Helicina costata Gray Jamaica
 calida Weinland Bahamas
 litoricola Gundl. Cuba
 bryanti Pf. Bahamas
 pfeifferiana Arango Cuba

(SCHRAMMIA)

Helicina conuloides Guppy (*schrammi* Crosse,)
    Dominica, Guadelupe

(PACHYSTOMA).

Helicina rostrata Morelet Guatemala
 bellula Gundl. Cuba
 rotunda Orb. ,,
 rhodostoma Gray Dominica
 caracolla Moricand Brazil
 occidentalis Guild St. Vincent
 ksatei Pféiff. Grenada
 oweniana Pf. Guatemala
 zephirina Duclos Mexico, Centr. America
 nemoralis Guppy Trinidad
 columbiana Phil. Venezuela
 neritella Lam. Jamaica
 jamaicensis Sow ,,
 ampliata Adams ,,

(OLIGYRA).

Helicina fasciata Lam. Guadelupe, &c.
 guadelupensis Sow. Martinique, &c.
 antillarum Sow. (*mazei* Crosse) Martinique
 moussoniana Pf. Bahamas
 reeveana Pf. Cuba
 rawsoni Pf. Bahamas
 foveata Pf. St. Thomas
 epistilia Guppy St. Lucia, Dominica

Helicina velutina Guppy   Dominica
    var. *guppyi* Pease (*humilis* Guppy) Dominica
     barbata Guppy   Trinidad
     substriata Gray   Barbados
     occulta Say   Virginia
     convexa Pf.   Bermuda
     phasianella Sow.   Haiti
     sericea Drouet   Venezuela
     orbiculata Say   N. America
     platychila Muhl.   Martinique
     subglobulosa Poey   Florida
     striatula Sow.   Martinique]
     dubiosa Adams (Alcadia) Jamaica
     megastoma Adams   „   „

(IDESA).

Helicina primeana Gass.   New Caledonia
     litoralis Montr.   „
     borneensis Pf.   Borneo

(EMODA)

Helicina porphyrostoma Gass.   New Caledonia
     spheroidea Pf.   Loyalty I.

Subgenus ALCADIA

(ISOLTIA)

Helicina nuda Arango   Cuba

(VIANA)

Helicina sagraiana Orb.   Cuba
    var. *ochracea* Poey   „

(ALCADIA s.s.)

Helicina brownei Gray   Jamaica
     hollandi Adams   „
     palliata Adams   „
    var. *citrinolabris* Adams   „
    var. *consanguinea* Adams   „
     major Gray   „
     microstoma Adams   „

GENUS PROSERPINA GRAY

Proserpina depressa Orb.	Cuba
linguifera Jonas	Jamaica
var. *pulchra* Adams	,,
nitida Gray	,,

GENUS CERES GRAY

(*Ceres colina* Duclos)

THURSDAY, 9TH AUGUST, 1894.

S. Devenish, M.A., Vice-President, in the Chair.

The Secretary read a notice of a paper on the Mammals of Trinidad by J. A. Allen and F. M. Chapman ; a notice of a paper on the Birds of Trinidad ; and a notice of a paper on the Orthoptera of S. Vincent and Grenada by C. B. von Wattenwyl.

THE MAMMALS OF TRINIDAD.

The Bulletin of the American Museum of Natural History, Vol. V. (1893) pp. 203-234, contains a paper by J. A. Allen and Frank M. Chapman on a collection of Mammals of Trinidad. To this paper is appended a complete list of the land mammals recorded for Trinidad and this list is now reproduced for convenience of reference. The species added by Messrs. Allen and Chapman are distinguished by an asterisk.

Cebidæ.

Mycetes (seniculus Linn.) Red Howler

Cebus sp. Sapajou, Capuchin Monkey

Vespertilionidæ.

Vespertilio nigricans Wied.

Thyroptera tricolor Spix

Emballoneuridæ.

Furipterus horreus Cuv.

Saccopteryx bilineata Temm.

　leptura Schreber

　canina Wied.

Rhynchonycteris naso Wied.

Noctilio leporinus Linn.   Fish-eating Bat.

Molossus rufus Geoffr.

　obscurus Geoffr

Phyllostomatidæ.

Chilonycteris rubiginosa Wagn.

Pteronotus davyi Gray

Mormops megallophylla Peters

Lonchorina aurita Tomes

Micronycteris megalotis Gray

Phyllostoma hastatum Pall.

Hemiderma brevicaudum Wied.

Glossophaga soricina Pall.

Anoura geoffroyi Gary

○ Chæronycteris intermedia All. & Chapm.

Artibeus perspicillatus Linn.

    planirostris Spix

    hartii Thomas

    quadrivittatus Peters

Vampyrops caraccioli Thomas

Chiroderma villosum Peters

Sturnira lilium Geoffr.

Desmodus rufus Wied. Bloodsucking Bat.

Felidæ.

Felis sp. Ocelot—Tiger Cat.

Mustelidæ.

Galictis barbara Linn. Wood-dog

Lutra insularis Cuv. Otter

Procyonidæ.

Procyon cancrivorus Cuv. Mangrove-dog

Cercoleptes caudivolvus Pall. Kinkajou

Sciuridæ

Sciurus æstuans Peters   Squirrel

Muridæ

Holochilus squamipes Brants

* Nectomys palmipes All. & Chapm.

* Tylomys couesii All. & Chapm.

* Orizomys speciosus All. & Chapm.

*   trinitatis All. & Chapm.

*   velutinus All. & Chapm.

*   brevicaudus All. & Chapm.

* Abrothrix caliginosus Tomes

* Mus rattus Linn. Black Rat

*   alexandrinus Geoffr. Roof Rat

    musculus Linn. House Mouse

Heteromyidœ

Heteromys anomalus Thompson.   Pouched Rat

Octodontidœ

Loncheres guianæ Thomas.   Spiny Rat
*   castaneus All. & Chapm.   Aguti Rat
* Echimys trinitatis All. & Chapm.   Pilori

Hystricidœ

Synetheres prehensilis Linn.   Tree Porcupine

Dasyproctidœ

Dasyprocta aguti Linn.   Aguti
Cœlogenys paca Linn.   Lap.

Cervidœ

Cariacus nemorivagus Cuv.   Deer—Biche

Dicotylidœ

Dicotyles tajacu Linn.   Collared Peccary.   Quenk
        labiatus Cuv.   Whitelipped Peccary.   Quenk

Bradypodidœ

Cholæpus didactylus Linn.   Two-toed Sloth

Myrmecophagidœ

Myrmecophaga jubata Linn.   Great Anteater
Tamandua tetradactyla Linn.   Tamandua
Cyclothurus didactylus Linn.   Little Anteater—Poor-me-
one

Dasypodidœ

Tatusia novemcincta Linn.   Armadillo—Tatu

Didelphidœ

Didelphis marsupialis Linn.   Manicou
        philander Linn.   Manicou gros-yeux
◦   murina Linn.   Manicou gros-yeux

N.B.—Though not recorded hitherto in scientific publications the house
mouse and the black rat have been well-known inhabitants of the island
ever since its settlement by Europeans.

In 1886 and again in 1891 specimens of the red-bellied Squirrel were
received at the Gardens of the Zoological Society of London from
Trinidad. This species was determined as *Sciurus variegatus*.

THE BIRDS OF TRINIDAD.

Mr. Frank M. Chapman has published in the Bulletin of the American Museum of Natural History, Vol. VI. (1894) pp. 1–86, an article on the Birds of Trinidad. This is an interesting and valuable contribution to the ornithological branch of our local natural history. To render more accessible the results of Mr. Chapman's work so far as regards the nomenclature of the specimens in the Leotaud Collection in our museum I append the list given by him and the numbers attached to the specimens.

Captain Kelsall who was in Trinidad in 1868 had a list of the Birds printed and in this list the Species named in Leotaud's book but not represented in his collection were indicated. This list included also the name [*] of the Scarlet Tanager not mentioned in Leotaud's book and on the occurrence of which in Trinidad a note was communicated to the Scientific Association by Captain Kelsall (see Proc. Scient. Assoc. Trinidad, 1868 p. 208). As the list prepared by Capt. Kelsall is out of print that which follows may be useful. It includes not only the French vernacular names given by Leotaud, but the English ones given by Mr. Chapman.

The present list will not only serve as a check-list of the Birds of Trinidad but as an index to the collection in the Victoria Institute. Notwithstanding the alterations and corrections made by Mr. Chapman the paper of the latter is in itself testimony to the very excellent work of Leotaud.

---

[*] I observe that Mr. Chapman regards *Piranga (Tanagra) aestiva* as synonymous with *P. rubra* Linn. Kelsall gives it as a different species, citing *P. erythropis* Vieill, as a synonym.

The number of Bird Forms worked out and identified by Leotaud was 297, raised to 298 by Kelsall. Mr. Chapman gives 303 names, besides three of Leotaud's not identified by him, making 306 in all. Mr. Chapman records 12 species not mentioned by Leotaud. In four cases he fuses under one name two forms given separately by Leotaud.

R. J. L. G.

LIST OF THE BIRDS.

In the following list the species marked with an asterisk * are unrepresented in the Leotaud collection.

Merula flavipes Vieill. Yellow-footed Thrush. Grive à pattes jaunes. 104
    (Turdus flavipes Leotaud.)

—— phæopygus Cab. Whitethroated Thrush. Grive à cravatte 103
    (Turdus phæopigus Leotaud)

—— xanthoscelus Jard. Blackbird. Grive noire ... 105
    (Turdus xanthoscelus Leot.)

Merula fumigata Licht. Cacao Thrush. Grive des Cacaos ... 107
    (Turdus casius Leot.)

—— gymnopthalma Cab. Barecheeked Thrush. Grive à paupieres jaunes 106
    (Turdus nudigenus Leot.)

Troglodytes rufulus Cab. Godbird. Rossignol 88
    (Troglodytes tobagensis Lawr.)

Thryothorus rutilus Vieill. Bush Wren. Rossignol des Halliers 89
    (Troglodytes rutilus Leotaud.)

Protonotaria citrea Bodd. Prothonotary Warbler. Fauvette à tête jaune 92
    (Mniotilta citrea Leotaud.)

Compsothlypis pitiayumi Vieill. Golden Sucrier. Sucrier doré 94
    (Mniotilta venusta Leotaud.)

Dendroica æstiva Gmel. Canary. Figuier 91
    (Mniotilta petechia Leotaud.)

Seiurus noveboracensis Gmel. Water Thrush. Batte-queue 90
 (Enicocichla noveboracensis Leotaud.)
Geothlypis æquinoctialis Gmel. Manicou ... 95
 (Trichas velatus Leotaud.)
Setophaga ruticilla Linn. Redstart. Officier 133
Basileuterus vernivorus Vieill. Fauvette des Halliers ... 96
 (Trichas bivittatus Leotaud.)
Cœreba luteola Cab. Sucrier 61
 (Certhiola flaveola Leotaud.)
Arbelorhina cærulea Linn. Green-legged Grampo. Grim-
 pereau à pattes soufre 58
 (Cœreba cærulea Leotaud.)
Arbelorhina cyanea Linn. Red-legged Grampo. Grimpereau
 à pattes roses 57
 (Cœreba cyanea Leotaud.)
Chlorophanes spiza Linn. Blackheaded green Honey sucker
 Vertvert à tête noire 59
 (Dacnis spiza Leotaud.)
Dacnis cayana Linn. Verdigree. Vert de gris 60
— plumbea Lath. Sucrier des Mangles 93
 (Mniotilta bicolor Leotaud.)
Cyclorhis flavipectus Sclater. Piegrièche 141
Vireo chiviagilis Licht. Petit Siffleur à tête grise ... 134
 (V. olivaceus Leotaud).
—— calidris Linn. Grand Siffleur à tête grise 134
 (V. altiloquus Leotaud.)
Hylophilus aurantiifrons Lawr. Petit Gobemouche ... 97
 (H. insularis Leotaud).
Progne chalybea Gmel. Martin. Hirondelle noire ... 45
 (Progne purpurea Leotaud)
Atticora cyanoleuca Vieill. Swallow. Hirondelle à ventre
 blanc 43
 (Hirundo cyanoleuca Leotaud)
Tachycineta albiventris Bodd. Swallow. Hirondelle à dos
 vert. 44
 (Hirundo albiventer Leotaud.)

Chelidon erythrogaster Bodd. Barn Swallow. Hirondelle
  à ventre roux 42
  (Hirundo rufa Leotaud.)
Stelgidopteryx uropygialis Lawr. Swallow. Hirondelle à
  ventre jaune 46
  (Cotyle uropygialis Leotaud.)
Procnias viridis Ill. Blue Mantle. Cottinga bleue   ... 138
  (Tersa ventralis Leotaud.)
Euphonia violacea Linn. Louis d'or Simple   167
—— trinitatis Strickl. Cravat. Louis d'or à cravatte   ... 168
  (Euphonia chlorotica Leotaud.)
—— nigricollis Vieill. Louis d'or à tête bleue 169
  (E. aureata Leotaud.)
Calliste desmaresti Gray. Worthless. Vertvert a tête caco. 164
—— vieilloti Scl. Variegated Tanager. Diable enrhumé ... 165
—— guttata Cab. Tiger Tanager. Arrivant   166
Tanagra cana Tayl. Bluebird. Oiseau bleu   158
  (T. glauca Leotaud.)
—— palmarum Ridgw. Palmist   159
  (T. olivascens Leotaud.)
——subcinerea Scl. Blueheaded Tanager. Grosbec à tête bleue 160
Ramphocelus jacapa Lafr. Silverbeak. Bec d'argent   ... 155
Piranga rubra Linn. Summer Tanager. Cottinga rose. ⎱ ?156
  (Pyranga aestiva Leotaud) ⎰ 298
—— hæmalea S. & G. Rufous Tanager. Cardinal a gros-bec 157
  (Pyranga hepatica Leotaud.)
Phænicothraupis rubra Vieill. Cardinal   161
  (Tachyphonus ruber Leotaud.)
Lanio laurencei Sclater   ...   . *
Tachyphonus luctuosus Lafr. Little Parson. Petit Père noir 163
  (Tachyphonus albispecularis Leotaud.)
—— rufus Bodd. Parson. Père noir...   162
  (T. beauperthuyi Leot.)
Chlorospingus leotaudi Chapman   ...   . *
Saltator albicollis Vieill. Gros-bec tacheté   154
  (S. striatipectus Leot.)

(Mellisuga o. Leot.)

Calliphlox amethystina Gmel. Amethyst 73

(Calathorax enicurus Leot.)

Chrysolampis mosquitus Linn. Ruby-topaz. Rubis-topaze... 74

(Mellisuga moschita Leot.)

Petasophora delphinae Less. Blue-eared Hummer. Colibri
à oreilles 66

(Polytmus d. Leot.)

Floricola longirostris Vieill. Carmine. Gorge carmin ... 75

(Mellisuga l. Leot.)

Agyrtria chionopectus Gould. Whitebreast. Colibri à
gorge blanche 71

(Polytmus ch. Leot.)

Polytmus thaumatias Linn. Pearl. Vert-perlé ... 67

(P. viridis Leot.)

Amazilia erythronota Less. Emerald. Raimonpe ... 68

(Polytmus e. Leot.)

Eucephala cærulea Vieill. Saphir 77

(Hylocharis c. Leot.)

Chlorostilbon atala Less. ... ... ○

Panyptila cayanensis Gmel, Swallow. Hirondelle à gorge
blanc 37

(Cypselus cayanensis Leot.)

Chætura cinereiventris Ridg. Rainbat. Petit Hirondelle
à croupion gris 39

(Acanthylis oxura Leot.)

——spinicauda Temm. Rainbat. Hirondelle à croupion gris 40

(Acanthylis poliourus Leot.)

——polioura Temm. Rainbat. Hirondelle à croupion gris ○

Hemiprocne zonaris Shaw. Ringed Growrie. Hirondelle
à collier blanc 38

(Acanthylis collaris Leot.)

Cypeloides rutilus Vieill. Swallow. Hirondelle à collier roux 41

(Hirundo r. Leot.)

Chordeiles acutipennis Bodd. Nighthawk. Engoulevent
à queue fourchue 35

(Ch. minor Leot.)

Sarcoramphus papa Linn.   King Corbeau.   King Vulture.
Roi des Corbeaux   1
Cathartes aura Linn. Cedros Vulture.   Corbeau à tête rouge   2
Catharista atrata Bartr.   Town Vulture.   Corbeau   ...   3
    (Cathartes fœtens Leot.)
Ictinia plumbea Gmel.   Plumbeous Kite.   Gabilan bleue...   22
Elanoides forficatus Linn.   Scissor-tailed Kite.   Queue-en-
    ciseaux   16
    (Mauclerus furcatus Leot.)
Rostrhamus sociabilis Vieill.   Hookbilled Kite.   Gabilan à
    bec crochu   17
    (R. hamatus Leot.)
Circus maculosus Vieill.   Harrier.   Gabilan à longue queue   25
    (C. macropterus Leot.)
Buteo abbreviatus Cab.   Small black Buzzard.   Petit Gabilan
    noir   5
    (B. zonocercus Leot.)
Urubitinga urubitinga Gmel.   Eaglehawk.   Gros Gabilan
    noir   8
    (Morphnus u. Leot.)
—— anthracina Licht.   Black Hawk.   Gabilan noir   ...   23
    (Astur unicinctus Leot.)
—— albicollis Lath.   Gabilan à dos noir   ...   4
    (Butes poecilinotus Leot.)
Asturina nitida Lath.   Speckled Hawk.   Gabilan ginga ...   24
Harpagus bidentatus Lath.   Toothed Falcon.   Gabilan à
    deuxdents   15
Gampsonyx swainsoni Vig.   Brown Hawk.   Grigri   ...   21
Leptodon unicinctus Temm.   Gabilan bleuâtre ...   19 & 20
    (Cymindis unicinctus and C. pucheradi Leot.)
—— cayenensis Gmel.   Guiana Hawk.   Gabilan à tête bleue   18
    (Cymindis cayanensis Leot.)
Spizætus mauduyti Daud.   Crested Spizætus.   Gabilan à
    huppe   6
    (Spizætus ornatus Leot.)
—— tyrannus Wied.   Speckled-leg Spizætus.   Gabilan à
    pattes ginga   7
    (Sp. braccata Leot.)

Ægialitis semipalmata Bon. Ringnecked Plover. Petit
Collier 208
(Charadrius s. Leot.)

Charadrius dominicus Mull. Golden Plover. Pluvier doré 209
(Ch. virginicus Leot.)

—— squatarola Linn. Black-bellied Plover. Gros Pluvier
doré 206
(Squatarola helvetica Leot.)

Numenius borealis Lath. Eskimo Curlew. Petit Bec
crochu 231

—— hudsonicus Lath. Hudsonian Curlew. Bec crochu... 230

Actitis macularia Linn. Spotted Sandpiper. Ricuit 239 & 240
(Tringoides hypoleuca and macularia Leot.)

Tryngites ruficollis Vieill. Buff-breasted Sandpiper. Petit
pieds jaunes 245
(Tringa rufescens Leot.)

Bartramia longicauda Bechst. Bartramian Sandpiper.
Pieds jaunes à longue queue 241
(Tringoides bartramius Leot.)

Symphemia semipalmata Gmel. Willet. Ailes blanches... 238
(Totanus s. Leot.)

Totanus solitarius Wilson. Solitary Sandpiper. Grandes
Ailes 235
(T. chloropygius Leot.)

—— flavipes Gmel. Yellow legs. Pieds jaunes. ... 236

—— melanoleucus Gmel. Greater Yellow-legs. Clin-clin 237

Limosa hæmastica Linn. Hudsonian Godwit. Bécard
ailes blanches ... ... 233 & 234
(L. hudsonica and ægocephala Leot.)

—— fedoa Linn. Marbled Godwit. Grand Bécard ... 232

Calidris arenaria Linn. Sanderling. Bécasse blanche ... 251

Ereunctus occidentalis Lawr. Western Sandpiper. Bécasse
à long bec 250
(Heteropoda longirostris and H. mauri Leot.)

Ereunctus pusillus Linn. Semipalmated Sandpiper. Bécasse
ordinaire 249
(Heteropoda semipalmata Leot.)

Erismatura dominica Linn. Squat Duck. Vingeon   ... 275
Aythya affinis Eyt. Lesser Scaup. Canard france   ... 273
   (Fuligula marila Leot)
Spatula clypeata Linn. Shoveller. Canard Spatule   ... 271
Anas discors Linn. Bluewinged Teal. Sarcelle à croissants 270
   (Pterocyanea discors Leot.)
—— americana Gmel. American Widgeon. Vingeon... 268
   (Mareca a Leot.)
Fregata aquila Linn. Man of War or Frigate Bird. Fregate 294
   (Atagen aquila Leot.)
Pelecanus fuscus Linn. Brown Pelican. Grand gosier   ... 293
Phalacrocorax brasiliensis Gmel. Cormorant. Plongeon
   à bec crochu *292
   (Graculus carbo Leot)
, Plotus anhinga Linn. Anhinga. Darter. Plongeon-soie... 289
Sula leucogastra Bodd. Black-and-white Booby. Fon
   commun   290
   (S. parva Leot.)
——piscator Linn. White Booby. Fou à pattes rouges   ... 291
Rhynchops nigra Linn. Black Skimmer. Bec-en-ciseaux... 280
Anous stolidus Linn. Noddy. Mauve noire *288
   (Anous melanogenys Leot.)
Sterna antillarum Less. Least Tern. Petite mauve   ... 287
   (Sterna argeutea Leot.
—— dougalli Mont. Roseate Tern. Mauve à bec noir ... 283
   (St. paradisea Leot.)
—— eurygnatha Saunders. Black-legged Tern. Grande
   Mauve à pattes noires   ...   .   * 285
   (St. elegans Leot.)
—— maxima Bodd. Royal Tern. Mauve à queue
   blanche   ...   . 281 & 286
   (St. cayenensis and regia * Leot.)
Phœthusa magnirostris Licht. Yellow-footed Tern. Mauve
   à patte jaune soufre   282
   (St. chloripoda Leot.)
Geochelidon nilotica Hass. Marsh Tern. Mauve à dos
   cendré   284
   (Sterna aranea Leot.)

THE ORTHOPTERA OF ST. VINCENT AND GRENADA
W.I.

A valuable series of papers on the Fauna of the West Indian Islands has been for some time in course of publication in English Scientific Journals. They are the result of the labors of a joint Committee of the Royal Society of London and the British Association. It is desirable for convenience of reference that some record of these papers should be included in the publications of this Institute and I therefore append hereto a list of the Orthoptera of the Islands of St. Vincent and Grenada taken from papers by C. Brunner Von Wattenwyl published in the Proceedings of the Zoological Society of London, 1892, p. 196 and 1893, p. 599.

		St. Vincent	Grenada
Dermaptera.			
Labia arcuata Scud.	...	x	x
rotundata Scud.	...	x	x
brunnea Scud.	...	x	x
pulchella Serv.	...	x	
Psalis americana Pal.	...		x
Spongophora sp.	...		x
Anisolabis janeirensis Dohrn	...	x	
maritima Bon.	...	x	x
Blattodea			
Anaplecta lateralis Burm.	...		x
Chorisoneura mysteca Sauss.	...		x
Anaptycta bipunctulata Brunn.	...	x	x
Phyllodromia adspersicollis Stal.	...	x	x
delicatula Guer.	...		x
notata Brunn.	...		x

Species		
Pseudophyllodromia semivitrea Brunn.	x	
albinervis Brunn.		x
Ischnoptera occidentalis Sauss.		x
Pelmatosilpha marginalis Brunn.		x
Periplanetæ australasiæ Fabr.		x
Epilampra brevis Burm.	x	x
Homalopteryx laminata Brunn.	x	
Stilopyga antillarum Brunn.	x	x
Panchlora viridis Burm.	x	x
Leucophæa surinamensis Linn.	x	x
maderæ Fabr.	x	x
Nauphœta lævigate Pall.		x
Holocompsa collaris Burm.	x	
Parasphærianigra Brunn.	x	
rufipes Brunn.		x
Latnidia castanæ Brunn.		x
Mantoœdea.		
Musonia surinama Sauss.	x	x
Parastigmatoptera lobipes Brunn.	x	x
Phasmodea.		
Phanoceles curvipes Brunn.	x	
Bacteria cyphus Westw.	x	x
linearis Drury	x	x
Diapherodes gigas Drury	x	x
Acridioidea		
Orphula punctata Geer	x	x
Tettix quadriundulatus Brunn.	x	x
Vilerna æneoculata Geer	x	x
Caletes apterus Brunn	x	x
Schistocerca pallens Thunb.	x	x
columbina Thunb.	x	x
Osmilia cœlestis Burm.		x
Anaulacomera antillarum Brunn.		x
laticaudata Brunn.	x	
Microcentrum pallidum Brunn.	x	x
Stilpnochlora marginella Serv.	x	x

Bliastes superbus Brunn. ...	x	x
Striolatus Brunn. ...	x	x
Cyrtophyllus crepitans Brunn ...	x	
Copiophora brevicornis Redt. ...	x	
Conocephalus guttatus Serv. ...	x	x
muticus Redt. ...	x	x
maxillosus Fabr. ...	x	x
infuscatus Scud. ...	x	x
frater Redt. ...	x	
heteropus Bol. ...	x	
macropterus Redt. ...	x	
punctipes Redt. ...	x	x
surinamensis Redt. ...	x	x
Xiphidium saltator Sauss. ...	x	x
propinquum Redt. ...	x	x
Pherterus cubensis Haan ...	x	x
Gryllodea		
Gryollatalpa hexadactyla Perty ...	x	x
Scapteriscus didactylus Latr. ...	x	
Tridactylus minutus Scud. ...	x	x
Amirogryllus muticus Geer. ...	x	
Nemobius cubensis Sauss. ...		x
Paragryllus rex Sauss. ..		x
Gryllus assimilis Fabr. ...	x	x
Gryllodes rufipes Brunr.. ...	x	
Ectatoderus antillarum Brunn. ...	x	x
Larandus marmoratus Brunn. ...	x	
Paræcanthus sp. ...		x
Endacustes dispar Brunn. ...	x	
Cyrtoxiphus vittatus Bol. ...	x	x
gundlachi Sauss ...	x	
Orocharis gryllodes Pall. ...	x	x
Metrypus luridus Sauss ...	x	x
claudicaus Brunn. ...		x
heros Brunn.	...	x
Podoscirtus modestus Brunn. ...	...	x

BIBLIOGRAPHICAL NOTICES.

The Zoological Proceedings for 1893 also contain (p. 692) a paper by G. W. and E. G. Peckham on the Spiders of the family Attidæ of the Island of S. Vincent ; and (p. 705) a paper by P. R. Uhler on the Hemiptera-heteroptera of the same Island. The same author contributes a further paper on the same subject to the Zoological Proceedings for 1894 (p. 156). Many new species are described. But descriptions of species without figures are, to say the least, unsatisfactory.

It may be useful also to record the titles of the following papers published during the past few years on the Natural History, Geology, &c., of Trinidad and its neighborhood.

G. A. Boulenger.—On Reptiles, Batrachians and Fishes from the Lesser West Indies. P Z S. 1891, p. 351.

R. J. L. Guppy.—On a Specimen of Pleurotomaria from Tobago, West Indies. P Z S. 1891, p. 484.

R. R. Mole and F. W. Urich.—Notes on some Reptiles from Trinidad. P Z S. 1891, p. 447.

E. Simon.—On the Spiders of the Island of S. Vincent. P Z S. 1891, p. 549 ; and 1894, p. 519.

R. J. L. Guppy.—Note on Bulimus oblongus. P Z S. 1892, p. 271.

H. Crosse.—Faune Malacologique terrestre et fluviatile de l'Ile de la Trinité. Journal de Conchyliologie 1890, p. 35.

A. J. Jukes-Browne and J. B. Harrison.— The Geology of Barbados. Journal of the Geological Society, May 1891 (Part I.) and May 1892 (Part II.)

R. J. L. Guppy.—The Tertiary Microzoic Formations of Trinidad. Journal of the Geological Society, November 1892.

R. J. L. Guppy.—The Microzoa of the Tertiary and other Rocks of Trinidad. Journal Field Naturalists' Club, 1893.

R. J. L. Guppy.—The Land and Freshwater Molluska of Trinidad. Journal of Conchology 1893, p. 210.

Lord Walsingham. The Microlepidoptera of the West Indies. P Z S. 1891, p. 492.

G. F. Angas. The Terrestrial Molluska of Dominica, P Z S. 1883, p. 594.

P. R. Uhler. The Hemiptera-Heteroptera of Grenada. P Z S. 1894, p. 167.

R. R. Mole & F. W. Urich. Biological Notes on some of the Ophidia of Trinidad. P Z S. 1894, p. 499.

In this paper the Authors give a preliminary list of the Ophidia recorded from the Island. This list includes the following names :—

Typhlops reticulatus Linn.
Glauconia albifrons Wagl.
Epicrates cenchris Linn.   Thicknecked Tree Boa.
Corallus cookii Gray.   Cascabel Dormillon.
Boa constrictor Linn.   Makauel.
      divinaloqua Laur.
Eunectes murinus Linn.   Huilia
Streptophorus atratus Hallow.
Geophis lineatus Dum. & Bib.   Ground Snake.
Liophis melanotus Shaw.   Beauty of the Road.
      reginæ Linn.
      cobella Linn.   Mapepire mang.
Coluber boddaerti Seetz.   Grass Machet.
      cornis Boie.   Cribo.
Spilotes variabilis Wied.   Tigre.
      pœcilostoma Wied.
Herpetodryas macropthalmus Jan.
      carinatus Linn.   ...   Machet.

Ahaetulla liocercus Wied.   ...   Lora
Petalognathus nebulatus Linn
Homalocranium melanocephalum Linn.
Oxybelis acuminatus Wied. ...   Liguis.
Dipsas cenchroa Linn   ...   Fiddlestring Mapepire.
Leptodira annulata Linn.
Scytale coronatum Schn.
Oxyrrhopus plumbeus Wied.
Elaps riisei Jan.   ...   Coral Snake.
    lemniscatus Linn.   ...   Large Coral Snake.
    corallinus Linn   ...   ...   Coral Snake.
Lachesis muta Linn.   ...   Mapepire Zanana.
Bothrops atrox Linn.   ...   ...   Valsain.

THURSDAY, 29TH NOVEMBER, 1894.

Sylvester Devenish, M.A., Vice-President, in the Chair.

The following Donations were announced :—

A bottle of Black Pepper grown at the Convict Depot, Chaguanas.—*C. W. Meaden.*

Smithsonian Report 1892.—*United States National Museum.*

Boletin del Instituto Geografico Argentino Tom xv. cuad. 1, 2, 3 & 4. *The Institute.*

Botanic Gardens—Bulletin No. 24.—*J. H. Hart.*

Report of Weather Bureau. U. S. 1893.—*J. H. Hart*

Summary do. do. do. —*J. H. Hart.*

Monthly Weather Review—May, June, July, 1894.—*J. H. Hart.*

Bulletin Kolonial Museum to Haarlem—March and July 1894.—*J. H. Hart.*

The following paper was read :—

OBSERVATIONS UPON THE PHYSICAL CONDITIONS AND FAUNA OF THE GULF OF PARIA.

BY R. J. LECHMERE GUPPY.

In 1887 I communicated to the Scientific Association of Trinidad what was intended to be the first part of a marine invertebrate Fauna of the Gulf of Paria. That part was confined to the Molluska. I had for many years previously studied partially the Crustacea and Echinodermata of the Gulf; but had failed to publish anything on these subjects owing partly to the incomplete state of my information and partly to the excessively absorbing and arduous nature of my official duties which left me little time or opportunity for natural history studies. Still I managed at intervals to do some work ; and among other things I discovered what then were new and undescribed species of crustacea, but could not follow up the subject. So much has since been done in the way of describing new species in this division of the animal kingdom especially in North America and in connection with the Voyage of the "Blake" and other national vessels in the West Indies that I think it likely that most of our new species are new no longer to science and nomenclature. Still in presenting some notes on the Gulf of Paria I will name such species of crustacea and echinodermata as have occurred to me and of which I have been able to identify the names, leaving it for future observations to complete correct and revise the list

In the first place I will begin these notes by a hasty glance at some of the physical conditions of the Gulf.

The Gulf of Paria lies in a depression caused by subsidence the long axis of which runs approximately E.N.E. and W.N.W. The extreme length of this basin is about 100 miles and its present depth below sea level nowhere exceeds 20 fathoms (except near the Bocas) and is generally from 10 to 15 fathoms. The estimated superficial area may as a rough approximation be stated at 3,000 square miles. Were this basin laid dry it would appear as a vast plain of mud nowhere rising into a perceptible elevation or hillock except in a few spots near the existing shores. Elsewhere I describe some characteristics of these shores; but I may now state that with the exception above mentioned the deepest water close off shore is to be found at Pointapier.

The Bocas are narrow channels connecting the Gulf of Paria with the Caribean Sea. These channels have had their origin as subaerial valleys, now submerged. (See papers in Proceedings Scientific Association, 1877, p. 103, and Agricultural Record 1891, p. 79). The first Boca (Boca de Monos) is barely half a mile wide, the second (Boca Huevos) a little wider. The third (Boca Navios so called I suppose because ships do not use it) about the same. The Grand Boca is about six miles wide. These widths may be taken to be the clear widths in the narrowest parts. The rapid currents through these channels have kept them scoured out and the depth reaches 120 fathoms. This probably may be the extremo amount of the depression caused by the geological movements referred to in my papers. No mud or sand of any kind is deposited in these Bocas. The bottom is one of rocky inequalities whose existence is evidenced in the violent eddies and ripples of the so-called "remu," a phenomenon which occurs when the water outside tho Bocas is at its lowest and consequently the water surface gradient from tho middle parts of the Gulf the greatest. Outside the Gulf the Caribean Sea varies in depth from 20 to 60 fathoms except in the line of the great down-throw passing along the axis of the Grand Boca. Here the depth attains 120 fathoms and is probably not less than 90 in any part. This line of depression seems to be a terrestrial feature of some magnitude. Its extension northeastward from the Boca Grande between the islands of Grenada and Tobago is indicated by deep water marking the boundary between the volcanic region of the Antilles and the non-volcanic region to the south and east thereof.

The rise and fall of the tide is from 2½ feet at neaps to about 4 feet at springs. We see in this the cause of the violent currents running through the Bocas. The vast body of water contained within the Gulf say 3,000 square miles to a depth of 2½ feet to 4 feet is all impelled towards the Bocas by the fall of tide in the Caribean Sea. The outward currents thus caused are given in charts at 2 to 4 knots, but I think it probable that they sometimes exceed this very considerably. The statement is probably correct as an average, though the extreme force of the currents at particular times and in particular spots may possibly even attain a velocity of ten miles an hour. These currents exhibit a number of extraordinary "tide rips" which vary greatly at different tides. The phenomena of these tides are greatly modified by the state of the Venezuelan rivers, which pour their waters into the Gulf, and which at times of heavy flood contribute so large an amount of water to the Gulf thus replacing what had passed out through the Bocas on the ebb that the current of the flood is often very slight, while the ebb is correspondingly strong. The time of high water is considerably affected by these land floods.

Observations on the tides of the Gulf are very much needed and such observations ought to be set on foot by the Government in view of the public benefits which would result from an accurate knowledge of the tides. I have paid more or less attention to the tides for the last twenty-five years and have calculated them for the local almanacks; but want of proper and sufficient observations have prevented the attainment of the requisite accuracy. Some time ago I was applied to on the subject by the United States Consul here and though I gave what information I could it was very meagre and uncertain. Tidal observations could be easily carried out with the proper instruments at a few stations in the Gulf. I should be willing to undertake the work of establishing the stations and making the observations.

The mixture of a large quantity of river water with that of the Gulf modifies both its saltness and its color. These effects are sensible in the wet season even as far out as the islands of Tobago and Grenada and beyond them to a distance of more than 100 miles from the Bocas. This produces a corresponding effect upon the animal and vegetable life of the region. Though Foraminifera and Ostracoda have forms belonging to estuarine and tidal waters and are abundant as fossils in the rocks of Trinidad they seem to be absent from the Gulf. I have hitherto failed in detecting any Ostracoda whatever. Of Foraminifera I found a few examples at the Bocas and some have occurred to me in dredgings near Sanfernando, but I believe these to have been derived from the foraminiferal rocks of Naparima. The most abundant animals in the Gulf are certainly those belonging to the Subkingdom Articulata. In this respect the water presents a parallel with the land, for on the latter insects are the most abundant form of life, while in the former their marine representatives the Crustacea outnumber all other creatures.

We spent a month this year on the island of Gaspari. Our station was at the Western end of the island and in full view of the first Boca. Gaspari forms the southern boundary of Chaguaramas Harbor. It is an island near two miles long and is entirely composed of the compact blue limestone of which El Pato (absurdly called Patos or Goose Island), Punta Gorda (the proper orthography ; See Carr's Voyage, Proc. Scient. Assoc. Trinidad, Dec. 1869, p. 396), Carera's and the Five Islands as well as the Laventille Hills are also formed. This limestone which I take to be of carboniferous or devonian age is largely rifted and cavernous. Hence the malarial fevers which prevail on the limestone islands and localities in the wet season. The rain-water charged with organic substances descends through the crevices into cavities into which the sea water enters. This mixture of freshwater charged with organic particles with the salt water is everywhere in the tropics the cause of malaria. The localities on the schistose formations away from swamps seem always to be healthy, but most of the sheltered bays and coves are always more or less unhealthy from the existence of swamps at their upper ends. They are submerged valleys as for example Scotland Bay, Chacachacare Harbor, Anse Maho (more lately called Dehert Bay and now Grand Fonds Bay). The swamps at such places as Teteron Bay, Chaguaramas &c. are due to submergence of land owing to movements of depression. It is however scarcely correct I think to attribute the swamps as Kingsley does to that want of tide which as he says is the curse of the West Indies. It is largely due to the rising and falling of the tide that these swamps exist in the state we find them. Were it not for the tide they would for the most part either be non-existent or they would be ponds and in either case would not give out those malarious exhalations which arise when the mud is uncovered.

At Gaspari I made observations on the temperature of the air and the water. The minimum temperature of the air was 76° while the maximum was 89°. On some days the minimum

was 78° while the maximum did not rise above 82°. During the
same period the temperature at my house on the savana in
Port-of-Spain was minimum 70° maximum 90°. The temperature
of the surface water of the Gulf was constantly 82° during the
time of our stay. We thought we perceived at times the existence
of a colder underlayer of water, but I was unable to verify the
fact. In October 1881 when we were under quarantine at the
Five Islands I found the temperature of the Gulf water to be
84°. The temperature of the Maraval river water at the time
was 76°. For many years I have made occasional observations
of the temperature of the river water. These observations were
always taken early in the morning before the water had time to
become heated by the sun's rays and I never found the temperature below 74° or above 78°.

These observations would appear to show.

1° That the temperature of the air at the Gulf
Islands does not usually fall so low or rise so
high as on the main island of Trinidad.

2° That the temperature of the sea water is much
higher than that of the river water. Probably
the temperature of the Gulf is itself lower than
that of the Caribean Sea (which observations
show to be about 84° at this time of year) owing
to the influence of the Orinoco and other rivers
in pouring a quantity of water of a lower temperature into the Gulf.

Further I observe that a minimum temperature of 78°
frequently occurs during a period of 24 hours at Gaspari while
in Trinidad the highest minimum that has ever occurred to me
was 77°, and the lowest maximum 78°.

For comparison with the above observations on the temperature of the water I add the following

On the 10th November at Piedmont, Sanfernando (120 feet above sea level) the bath (an iron bath in a partially inclosed gallery or veranda) was filled with water from the well in the evening (say about 5½ p.m.) The next morning at 7 a.m. I took the temperature. That of the air was then 75°; the temperature of the water in the well was 76° and that of the water in the iron bath 74°. The water of the well being in communication with the whole body of under ground water retained its average temperature, while that in the bath was exposed to evaporation and so lost some of its heat during the night. Humboldt observes that the temperature of rivers in tropical South America is from 24° to 28° centigrade (75° to 83° F.)

In times gone by when I had the energy for making researches in Natural History I had not the opportunity. Although I constantly traversed and retraversed the length and breadth of this island in the discharge of my duty it was always at high pressure, and I very rarely found the opportunity of indulging in any of my favorite pursuits. Now when I have the leisure the energy is gone. I found that one haul of the dredge—at the utmost two—was as much as I could manage. And considering that out of every ten hauls in our gulf only one brings up anything good or rare it will be seen that the results of my one or two dredging expeditions did not amount to much. Dead shells of common species came up at every haul of course—some places furnished broken dead coral and worn decayed shells in abundance while others produced little else but sand or even stinking mud. During my stay at Gaspari I got no good haul whatever—but once a fair but dead specimen of *Thracia dissimilis* came up. A small dead but good specimen of a rare *Solarium* occurred on one occasion and I found several examples of a new form of *crepidula* which lives on the very common *cerithium caudatum*. The conditions of some parts of our Gulf, notably Chaguaramas Harbor, seem to be very favorable to the latter shell. Strange to say an example of a

Pteropod (*Hyalea longirostra* Les.) occurred in one of these dredgings. This is a species which lives on the surface of the sea and of which I had once found an example cast ashore at highwater mark.

My notices and lists are extremely incomplete and rudimentary; but they may yet have their value and at all events on some points they are a beginning. On the subject of the molluska I have only inserted a few remarks in this paper as I propose at a future meeting to give a separate paper on this branch alone, as it is one that I have more fully investigated than some of the others.

Small Fishes are not uncommonly brought up in the dredge. But it would scarcely be useful to treat of these apart from the general ichthyological fauna of the Gulf and I am not at present in a position to do this. I exhibit however, a few specimens among which you will observe the curious pipefish (*Syngnathus*) and the seahorse (*Hippocampus*).

The surface of the sea, apparently so clear when we look at it in the daytime, swarms with the most varied forms of life some of which manifest themselves to us at night by their phosphorescence. The organism called *noctiluca* is credited with the larger share in the phosphorescence of the sea, but numerous minute crustacea and other organisms certainly contribute a large quota.

Diatoms abound in the Gulf both floating in the surface water and growing upon marine plants &c. at the bottom, especially in quiet places where seaweed exists. In the surface water Coscinodiscus occurs with Dictyocha & Navicula, while the forms found at the bottom include Synedra Diatoma & Asterionella.

My acquaintance with Diatoms is slight, but the more it is extended the more I am satisfied that the pseudo-generic divisions are similar to those of the Radiolaria and do not represent natural genera. Many of these reputed genera are in fact but various forms of the same species. The labors of Parker, Rupert Jones Brady and Carpenter have redeemed the Foraminifera from a similar opprobrium and made the study of these minute beings a source of pleasure instead of the pain arising from the hopeless confusion and never-ending multiplication of useless and misleading epithets.

It would be highly interesting to have an account of the peculiar botanical features of Gaspari and the analogous islands. The soil of these islands is peculiarly adapted to certain vegetation while it is unsuitable to others. However as my botanical knowledge is unfortunately very slight I must confine myself to a very few remarks. The limestone furnishes by disintegration an extremely fine-grained red clay which when wet adheres to every object it touches. As a guess it may be stated that an ounce of this clay represents a ton of disintegrated limestone. In some parts there is a considerable thickness (many feet) of this clay and very much is washed away by every shower. It makes a fairly fertile soil in which many plants thrive. Guinea grass grows wild in places where it is not overrun with bush. Palms however do not relish the limestone or the red clay. Yet attempts have been made, all resulting of course in failure, to establish the coconut palm.

One of the most notable plants here is a Spurge (Euphorbia) which I should say was one of the most venomous plants in existence. Not only is it extremely poisonous as plants of its family usually are, but it is covered, leaves and stem, with sharp needles which sting very severely. It took me several days to recover the effects of having run against one of these plants.

LIST of Crustacea & Echinodermata collected in the Gulf Paria or on its Shores.

CRUSTACEA.

Leptopodia sagitaria Fabr.   Long-leg Crab.
Chorinus heros Herbst
Mithrax cornutus Sauss.   Horned Crab.
    pleuracanthus Stimps.   „
    minutus Sauss.   „
Pericera bicornuta Latr.   „
Lambrus crenulatus Sauss.   Folded Crab.
Cancer lobatus ? M. Edw.
Panopeus occidentalis Sauss.
Micippe ? sp.
Eriphia gonagra Fabr.
Lupea cribraria Lam.
    anceps Sauss.
Boscia dentata Latr.   Mountain Crab.
Cardisoma guanhumi Marg.   Land Crab.
    quadrata Sauss   „
Gecarcinus lateralis Frem.   „
    ruricolus Linn.   „
    sp. indet   „
Ocypoda rhombea Latr.   Box Crab.
Gelasimus vocans Linn.   Calling Crab.
Sesarma cinerea Box   Mangrove Crab.
    sp. indet.   „
Grapsus pictus Latr.   Shore Crab.
Calappa marmorata Fabr.   Marble Crab
Matuta victor Fabr.   Victor Crab.
Hepatus fasciatus Latr.   Liver Crab.
    sp. indet.   „
Dromia lator ? M. Edw.   Ball Crab
    globosa ? Lam.   „
Raninoides sp. indet.

Hippa emerita Linn.

Pagurus granulatus Olivier

Palinurus americanus M. Edw.   Lobster

Squilla mantis Rond.

    scabricauda Latr.

Gonodactylus chiragrus Fabr.

Alpheus heterochælis Say.   Snapping Prawn

Peneus setiferus Linn.   Prawn

Hippolyte sp.   Humpbacked Prawn

Pandalus gurneyi Stimps.   ,,

Lucifer sp. indet.

Talitrus sp. indet.   Sandhopper

Æga affinis M. Edw.   Fish Louse

ECHINODERMATA.

Ophiura appressa Say

Ophiolepis elegans Lutken

    sp. indet.

Asteroporpa annulata Lutken

Oreaster reticulatus Rond.

Astropecten polyacanthus Müll.

Luidia senegalensis Lam.

Echinocidaris punctulata Desm

Psammechinus excavatus Blainv

Echinometra acufera Blainv.

Melita testudinata Klein.

Encope grandis Agass.

THE MOLLUSKA OF THE GULF OF PARIA.

By R. J. Lechmere Guppy.

The Gulf of Paria is a large landlocked sheet of water lying between the island of Trinidad and the continent of South America. It is over a hundred miles in extent from east to west and about fifty miles from north to south. To the north it is separated from the Caribean Sea by the high and long promontory of Paria and connected with it by narrow passages called " Bocas del Drago " or Dragon's Mouths, commouly called locally " the Bocas." To the south the Gulf is hemmed in by the delta of the Orinoco. That delta is intersected by several large branches of the river and some of these discharge themselves into the Gulf whose level during the wet season is sensibly raised its waters discoloured and its saltness greatly diminished by the immense quantity of fresh water thus poured into it. During this season a rapid current runs out through the Bocas. The flowing tide has scarcely power to check this current, while during the ebb it runs like a mill-race. The pale brownish tinge given by the admixture of river water is perceptible as far as Grenada and Tobago, or eighty to ninety miles out. In the dry season there is an inward flow of the tide through the Bocas but even then the outward flow is much stronger than the flood. To the southward between the island of Trinidad and the delta of the Orinoco there is a channel called the Serpent's Mouth connecting the Gulf of Paria with the Atlantic Ocean.

Except to the north where the Parian Range on the main-land and its continuation on the island, the northern mountains of Trinidad, abut on its waters, the shores of the Gulf are in general low and the water deepens so gradually that the depth of a yard is only attained at a mile off shore and the bottom to depths unknown is formed of a soft and yielding mud. In some such localities the line of demarcation between land and water is vague and indefinite being formed by tidal (mangrove) swamps. In all such parts of the Gulf the fauna is of a peculiar type, quite lacking the characteristic forms of truly marine life. The littoral mollusks here are the oyster, the periwinkle and the neritine ; while on the muddy bottom below tidemarks *Pyrula melongena*, *Venus pectorina* and *Mytilus brasiliensis* are abundant. Where a mud-flat exists to which some degree of firmness is imparted by an admixture of sand we find numbers of the most delicious bivalve in the world, namely *Asaphis deflorata*, a larger species than its eastern congener and representative *Asaphis rugosa*.

On the steeper shores and on those which approach the main sea the molluska are more varied and comprize with few exceptions the forms found throughout the region. Even here, however, we trace the influence of surrounding conditions in the color of many of the shells. This is of a more lugubrious and uniform hue than in the antillian representatives of the same species. As particularly striking examples of this I may quote *Voluta musica* and *Fasciolaria tulipa*. Of the former the Trinidad examples are very dark and the lines and spots are of an intense black instead of the pinkish-brown of antillian specimens. Of the latter named shell Trinidad examples are usually of a uniform brown without any lines spots or markings whatever. The dark and monotonous tints of the molluska of the West Coast of America has been made the subject of a paper by Fischer in the Journal de Conchyliologie 1873 (p. 105). The only circumstance we can find in common between the two regions is the absence of coral reefs. In seas where coral reefs abound the shells are of varied and brilliant hues but whether it is the presence of the coral reefs that gives colour to the shells or whether the phenomena are merely concomitant consequences of the same cause is not clear. In our case we might have ascribed the fact to the effect of the waters of the Orinoco ; but such an explanation will not meet the case of the west coast of South America, where large rivers are absent,

On the eastern shore of Trinidad where the Atlantic surges thunder on a sandy shore we have an assemblage of molluska with its own peculiarities and quite distinct from that of any other locality. Most of the shells found there have however occurred as stragglers on the shores of the Gulf and so are included in my list. I may name as characteristic shells of these Atlantic beaches (among others) *Trigona mactroides* (of the shells of which in places almost the entire beach is formed), *Cytherea dione*, *Mactra alata* and *Donax striata* and *denticulata*. Univalves are rare; but the logs of wood washed on shore generally become the resort of Litorinas. *Natica marocana*, whose prey is the Donaces, is not uncommon.

All the species enumerated in the following list have been collected by myself in the Gulf of Paria or on the coasts adjoining. Though consisting for the most part of names only yet this list represents a very large amount of work, not only in collecting but in the study and identification of the shells and in assigning to them their proper names amid the heap of synonyms and inaccurate determinations with which unfortunately the science of malacology is burdened. In respect of nomenclature I have been very greatly assisted by Krebs' Catalogue of the West Indian Marine Shells; a work published for private circulation only, but most valuable for its suggestions as to synonyms. I have however allowed myself considerable liberty in the fusion of specific names allowed to stand in Krebs' list; for the author of that work while condemning the "sickly passion" which possesses some naturalists for describing so-called new species, still allowed too many doubtful appellations to crowd his pages. And there are many names in his list which I strongly suspect to be redundant, yet I do not touch them for want of certainty as to what species were in view by the authors of the names.

Although I have been engaged for a lifetime (for that portion at least of it not occupied by my official duties) on the study of natural history and particularly of Geology and

Conchology I dare not affirm that all the species herein enumerated are correctly named. Some rectifications will have to be made without doubt, some species added and perhaps a few fused together : not many indeed for I have devoted much labour to the reduction of superfluous nomenclature. In fact I have given almost as much time to the fusion of specific names as many naturalists have to the making of species out of mere individual variations. Of the latter tendency I can point to a striking example in the work of my late friend Otto A. L. Mörch "On the West Indian Scalidæ," where something like forty species are made out of two or three ; and moreover these forty pseudo-species are arranged in an imposing array of sections each with its technical character, usually of no more real value than the presence or absence of a beard among individuals of the human race for example.

A naturalist finds a group of beings for which he devises a name and in order to comply with established usage he must make out some technical characters upon which to frame a diagnosis, definition or description. Another naturalist finds that other members of the same group have not these technical characters or have not all of them. He considers himself called on, not to alter or improve the diagnosis, but to invent a new name and a new diagnosis. But technical characters do not constitute groups ; and it often happens that as judged by technical characters individuals of the same species are more different from each other than different species of the same genus or group. The differences between species are often difficult of expression in language, nor can they be made the subject of chemical or mathematical tests. The cabinet naturalist is often at a loss to find characters whereby to separate what are really distinct species while he finds no difficulty in noting characters whereby to separate a single species into several, or even into two or three genera. To relieve himself from the difficulties he labors under he seeks characters of a distinct and definite and if possible of a mathematical kind so that he can certainly say "This has such a character therefore it belongs to such a group;" and "This has not such a character therefore it does not belong to that group." But nature does not work in this way; and our only course is when we find a group of beings in nature to name and describe it as well as we can and to conform our systems as nearly as possible to nature and not to try to force nature into an appearance of conformity with our systems.

In the synonymy the names quoted may not in all cases have been originally intended for the species under which they here stand, but may have been incorrectly applied by some author. Owing to the close resemblance of some species of the eastern hemisphere to western ones it has often happened that one name has been used for two distinct shells; and I am not sure that we have yet in all cases avoided the errors traceable to this source of confusion.

The principle on which the authorities are cited is that the specific names are assigned to the author who first described the species and not to the one who first adopted a particular collocation of generic and specific names. This principle has been generally acted up to by conchologists; but it is to be regretted that many writers on natural science have followed a different course thereby introducing unnecessary disorder into nomenclature. The want of a definite enunciation of the rule may be somewhat to blame for this. I propose the following—"The author's name appended to a specific name is to be regarded as the authority for the specific name *only*." It is absurd to see species described by Linné or Lamarck gravely ascribed to some modern catalogue-compiler or system-maker merely because the latter chooses to employ generic or subgeneric designations of later invention. If the principle now advocated be clearly recognized it is quite unnecessary to use the abbreviation "sp." or to place the name of the authority in parenthesis; these are merely ways of trying to escape an imaginary and wholly superfluous difficulty, that of giving at once the authority for the specific name and of indicating that the author of the species was or was not the authority for a more or less arbitrary collocation of generic and specific names. But a reference to Linné's work will show indeed that he does not closely collocate the generic and specific names. He places the generic names as a caption and following it are the definitions of the species to each of which is prefixed the name of the species only. Each specific name then has a separate existence independent of the generic name.

Woodward has dealt with the subject of the employment of defective generic names on insufficient grounds in his Manual (pages 59 to 61) and Fischer has also some remarks on the same subject very well worthy of study in the introduction to the 1857 volume of the Journal de Conchyliologie. In the volume for 1866 of the same publication (p. 381) Crosse has also some remarks on the point, and I agree with him in all except as to the "honouring" of individuals. The object of scientific nomenclature is the advancement of science and the convenience of the student of nature and not the honouring of any individuals. But nomenclature would be in an ever-shifting state and worse confused even than it is if the name of the author of a specific epithet can be changed by everyone who has a fancy to place the species in a new genus of his own or any one else's invention.

On the subject of inventing new names for well-known groups already named (such for example as Pelecypoda for Conchifera) see Carter on Sponges, Annals and Magazine of Natural History, 4 ser. vol. xvi, p. 4. It would be hard to devise designations or definitions entirely without fault, and therefore where one in use is fairly satisfactory it is a useless innovation or rather an injury to science to introduce a new name. The whole method and practice of science is liable to be brought into disrepute by this and other unnecessary and objectionable practices which are unfortunately only too much in vogue.

The classification I have adopted is that I have used in my own cabinet. It is in the main based upon that of Woodward by far the best ever published. Some of my innovations may seem hazardous, but I have never departed from my model without the most substantial reasons. The systems devised by various conchological writers seem to be founded on the most arbitrary and artificial characters and I have found myself unable to adopt them. And although the arrangement of my list may not in every point meet with the approval of all conchologists it is the best my information allows me to frame, and I believe it is in some respects an improvement upon many in use.

Order OCTOPODA.
OCTOPUS Cuvier.

1. **O. americanus** Orb.

O. vulgaris, D'Orbigny, Moll. Cuba, vol. i, p. 11, pl. 1, f. 1.

I cannot assert that there exist any characters sufficiently marked for the separation of this West Indian Octopus from the widely-distributed form called *O. vulgaris*. On his plate quoted above D'Orbigny calls this var. *americanus* but in the text he does not give that name, but merely points out the differences.

Order DECAPODA.
LOLIGO Lamarck.

2. **L. plei** Blainv. D'Orb. Moll. Cuba, vol. i, p. 42.

SEPIOTEUTHIS Férussac.

3. **S. sepioidea** Blainv. D'Orb. Moll. Cuba, vol. i, p. 34.

It is scarcely necessary to make any remark upon these species of Kephalopoda, which are not uncommonly taken by the fishermen in the Gulf. They are used as food.

SPIRULA Lamarck.

4. **Sp. fragilis** Lam. D'Orb. Moll. Cuba, vol. i, p. 64.

Order PTEROPODA.
HYALEA Lamarck.

5. **H. longirostra** Les. D'Orb. Voy. Amer. Merid pl. vi, f. 6–10.

5a. **Styliola subula** Quoy & Gaimard.

Order TECTIBRANCHIATA.

NOTARCHUS Cuvier.

6. **N. laciniatus** Rüppell. Gray, Fig. Moll. pl. clii, f. 1.

ACLESIA Rang.

7. **Acl. sp.** indet.

APLYSIA Linn.

8. **'Apl. nigra** Orb., Voy. Amer. Merid. p. 209, pl. xviii, f. 1, 2.

9. **Apl. sp.** indet.

BULLA Klein.

10. **B. maculosa** Mart.

B. striata Orb. Moll. Cuba. vol. i, p. 122.

B. media Phil.

B. umbilicata Bolt.

B. ibyx Meusch.

We have for many years kept the name of *B. striata* for our shell. But that name properly belongs to the Mediterranean form which is quite distinct from the West Indian shell. *B. maculosa* is nearer to *B. ampulla*. The Caribean Sea furnishes us with another species, *B. solida* Gmel. which I have not found on our shores. It is almost if not quite identical with *B. australis* Quoy and is still more like *B. ampulla* than *B. maculosa* is. *B. striata* is fairly delineated in the "Index Testaceologicus"; it is therein ascribed to the West Indies, and this is no doubt the source of our error.

Order HOLOSTOMATA.

(Sub-order Pyramidellina.)

TURBONILLA Risso 1826.

11. **T. turris** Orb. Chenu, Man Conch., vol. i., f. 1287.

Chemnitzia turris Orb. Moll. Cuba, vol. i., p. 219, pl. xvi., f. 10-13.

,, pulchella Orb. l. c., p. 220, pl. xvi., f. 14-17.

,, ornata Orb. l. c., p. 221, pl. xvi., f. 18-21.

Chemnitzia modesta Orb. l. c., p. 222, pl. xvi.,
    f. 22-24.

„	levis Adams.	Contr. Conch. p. 73.
„	pusilla Adams.	Ibid.	p. 74.
„	exilis Adams.	Ibid.	p. 74.
„	substriata Adams.	Ibid.	p. 73.
„	puncta Adams.	Ibid.	p. 72.

Also Ch. flavocincta and subulata Adams.

All the forms named above (and many others) belong to
the one species, which is plentiful in the Pliocene of Trinidad and
occurs sparingly in the Gulf of Paria.

(*Suborder Naticina.*)

NATICA Adanson.

12. **N. marocana** Chemn. Wood, Ind. Test. Nerita 14.

N. marochiensis (?) Lam.

Common on sandy beaches. With a falling tide a furrow
may be seen on the beach at the end of which will be found the
Natica with a Donax wrapped in its foot leisurely imbibing the
juices through a neat round hole bored in the shell of the
bivalve.

13. **N. canrena** Linn. Wood, Ind. Test. Nerita 1.

N. alapapilionis Chemn. Chenu, Conch. vol. i.,
    f. 1163.

N. proxima Adams (young).

Order CALYPTREACEA.

CREPIDULA Lamarck.

14. **Cr. aculeata** Chemn. Chenu, Conch. vol. i.,
    f. 2355-56.

D'Orbigny, Voy. Amer. Merid. p. 464, pl. lviii.,
    f. 4, 5.

15. **Cr. porcellana** Lam. Cuvier, Regne An. pl.
    xlvii., f. 5.

· Cr. protea Orb., Moll. Cuba, vol. ii., p. 192, pl.
    xxiv., f. 30-33

15a. Crepidula cerithicola n. sp.

Smooth; usually very convex externally, almost hemispherical, somewhat irregular in figure, marked with radiating flames or interrupted stripes of reddish brown which are generally visible internally through the shining subtransparent shell. Diameter of a large example about 8 millimeters. Inhabits the outside of *Cerithium caudatum* Sow. which is abundant in the shallow water off Gaspari. It is probably a variety of *Cr. porcellana*; but it is a very marked form which has been modified in accordance with the exigencies of its station.

CALYPTREA Lamarck

16. **C. auriculata** Chemn. Reeve, C. I. Crucibulum 15.
C. cuvieri Desh in Cuv. Regne An. pl. xlviii., f. 4.

17. **C. uncinata** Reeve, C. I. Calyptrea 17.

TROCHITA Schumacher

18. **Tr. candeana** Orb., Moll. Cuba, vol. ii., p. 190 pl. xxiv., f. 28, 29.

PILEOPSIS Lamarck.

19. **P. antiquata** Linn.
Patella mitrula Gmelin.
Capulus lamellosus Chenu, Conch. vol. i., p. 329, f. 2377.

Order CÆCIDÆ.

CÆCUM Fleming.

20. **C. formulosum** Folin, Journ. Linn. Soc. vol. x. p. 258, pl. viii., f. 4.

21. **C. phronimum** Folin (C. regulare Carp.?)

A well-defined and simple natural genus, but which (alas!) has not escaped the rage for superfluous nomenclature, generically and specifically.

21a. Meioceras nitidum Stimpson

Order DOCOGLOSSA.

PATELLA Linné.

22. **P. elegans** Phil.

 P. candeana Orb., Moll. Cuba vol. ii., p. 199, pl.
  xxv., f. 1-3.

 P. pulcherrima Guilding.

 Tectura fascicularis Menke.

SCUTELLINA Gray.

23. **Sc. antillarum** Shuttl.

Siphonariidæ.

SIPHONARIA Sowerby.

24. **S. lineata** Orb., Moll. Cuba, vol. i., p. 232, pl. xvii.,
  f. 13-15.

 M. Crosse in Journ. Conch. 1890, p. 52. has remarked upon
the change of D'Orbigny's name *lineolata* to *lineata*. The credit
of the emendation is due to H. Krebs and not to me.

Order CHITONACEA.

CHITON Linné.

25. **Ch. rugosus** Gray. Reeve, C. I. Chiton 115.

 Ch. squalidus Adams.

 Ch. guildingi Reeve.

26. **Ch. segmentatus** Reeve. Chenu, Conch. vol. i.,
  f. 2861.

 Chætopleura jancirensis Gray.

 Ch. apparata = rufocostata = asper Shuttl.

27. **Ch. caribeorum** Carp = Ischnochiton jamaicensis Carp.

28. **Ch. marmoratus** Gmelin.

29. **Ch. squamosus** Linn.

 The Chitons make nocturnal excursions on limestone rocks
as much as fifteen feet above high water mark. Where not
gnawed by these mollusks the surface of the rock is rough and
jagged, but where it has undergone the rasping process it is
smoothed and covered with fine striæ.

30. **Chiton (Acanthochiton) spiculosus**
Reeve.

Order CERITHIACEA.

TURRITELLA Lamarck.

31. **T. imbricata** Linn. Woodward, Man. Moll. pl. ix, f. 1
Hanley, Ips. Linn. Conch. p. 344, pl. iii, f. 2 (palevar.)

CERITHIOPSIS Forbes.

32. **C. punctatum** Linn. (not Brug. nor Lam.)
C. emersoni Adams.
C. subulatum Montf.

33. **C. nigricinctum** Adams

33a. **Cerithiopsis tubercularis** Mont.
Bittium greeni Stimps.

BITTIUM Leach.

34. **B. nigricans** Stimpson.

CERITHIUM Brugnières.

35. **C. gibberulum** Adams.
C. columellare Orb., Moll. Cuba, vol. ii, p. 155,
pl. xxiii, f. 13-15.

Bittium varium Pfeiffer.

36. **C. caudatum** Sow.

37. **C. minimum** Gmel.
C. zonale Brug.   C. ferrugineum Say.
C. striatum Say=C. septemstriatum Say.
C. eriense Val. Kien. Coq. Viv. Cerithium, pl. xxiv, f. 1
C. nigrescens Menke   C. megasoma Adams.
C. variabile Adams.

RISSOINA Orbigny.

38. **R. bryeria** Mont. Krebs, Proc. Scient. Assoc.
Trin. 1873, p. 55.
R. catesbyana Orb. Moll. Cuba, pl. xii, f. 1-3.
R. minor Ad.
R. elegantissima Orb. Moll. Cuba, pl. xii, f. 27-29
R. multicostata Ad.

R. scalarioides Adams, Contr. Conch. p. 113, and
R. firmata Adams.

R. lactea Brown, Ill. Conch. p. 11, pl. viii, f. 77=
R. pusilla Brocchi.

R. scalarella Adams.

R. dubiosa Adams, Contr. Conch. p. 114=R.
scalarella Adams.

R. dunkeri Pfeiff., R. candida Brown and R.
chesneli Mich.

38a. **Rissoina browniana** Orb. Moll. Cuba vol.
ii, p. 28, pl. xii, f. 33–35.

R. lævissima Adams.

R. lævigata Adams.

R. sloaneana Orb. Moll. Cuba, vol. ii, p. 28, pl.
xii, f. 36–38.

R. vitrea Adams.

BARLEIA Clark.

39. **B. tincta** n. sp.

Conic-oblong, rather stout, of a horny pinkish-brown appearance; whorls 5–6 smooth; aperture suboval, almost circular slightly angulate above. Lip sharp, very slightly expanded. Operculum deep red.

I should have taken our shell to be a *Hydrobia* but for its operculum which appears to resemble that of *Barleia rubra* Mont. Indeed the shell itself is very like that one. It occurred among weed in 1–2 fathoms water in the Gulf.

From *B. congenita*, Smith Z. P. 1890, pl. xxi, f. 25 it differs in the less ventricose whorls.

Prof. W. H. Dall considers it to be *Assiminea (Paludestrina) affinis* Orb. (Moll. Cuba, vol. ii, p. 8, pl. x, f. 8.)

LITORINA Férussac.

40. **L. nodulosa** Gmel. Orb. Moll. Cuba, vol. i. p.
205, pl. xiv. f. 11–14.

L. dilatata Orb. l. c. p. 207, pl. xiv, f. 20–23.

L. tuberculata Orb. l. c. p. 206, pl. xiv, f. 15–19.

41. **L. muricata** Linn. Chenu, Conch. vol. i, f. 2112.

42. **L. flava** King. D'Orb. Voy. Amer. Merid. p. 391, pl. liii, f. 1–3.

      L. irrorata var. Petit.

43. **L. ziczac** Chemn. Chenu, Conch. vol. i, f. 2107.

      L. lineata Orb. Moll. Cuba, vol. i, p. 208, pl. xiv, f. 24–27.

44. **L. columellaris** Orb., Moll. Cuba, vol. i, p. 213, pl. xv. f. 18–20.

45. **L. tigrina** Orb., l. c. vol. i., p. 211, pl. xv., f. 9-11.

      L. undulata Orb. (not Phil.) Moll. Cuba, vol. i., p. 212, pl. xv., f. 12-14.

46. **L. angulifera** Lam. Chenu, Conch. vol. i., f. 2093.

      L. scabra Orb. (as of Linn.) Moll. Cuba, vol. i., p. 212, pl. xv., f. 15-17.

FOSSAR Adanson.

47. **F. trochlearis** Adams.

MODULUS Gray.

48. **M. lenticularis** Chemn. Chenu, Conch. vol. i., f. 2122.

49. **M. unidens** Chemn. Wood, Ind. Test. Trochus 71.

      Trochus perlatus Wood.

      Monodonta carchedonius Lam.

SOLARIUM Lamarck.

50. **S. tesselatum** Desh. Wood, Ind. Test. pl. xxix. f. 56.

VERMETUS (Adanson).

51. **V. conicus** Dilw.

This name is inserted merely to indicate the presence of the genus in the locality. Whether there are more species than one and whether this is the right name of the species (or of one of them) I am unable to aver with positive certainty.

SIPHONIUM Browne.

52. **S. decussatum** Gmel. Wood, Ind. Test. Ser
pula 30.

Chenu, Conch. vol. i., f. 2303.

Order VOLUTACEA.

VOLUTA Linn.

53 **V. musica** Linn. Woodward, Man. Moll. pl.
vii., f. 9.

MARGINELLA Lamarck.

54. **M. cœrulescens** Lam. Wood, Ind. Test. Voluta 68.

M. subcœrulea Mart.

Voluta prunum Gmel.

55. **M. marginata** Born. D'Orbiguy, Moll. Cuba,
vol. ii., p. 96.

M. cincta Kien.

M. bivaricosa Lam. Chenu, Conch. vol. i.
f. 1057-8.

M. interrupta (Lam.) D'Orbigny, Moll. Cuba, vol.
ii., p. 97.

56. **M. imbricata** Hinds.

CYPRÆA Linné.

57. **C. exanthema** Linn. Chenu, Conch. vol. i., f. 1675.

Order CONACEA.

CONUS Linné.

58. **C. testudinarius** Mart.

PLEUROTOMA Lamarck.

59. **Pl. antillarum** Orb., Moll. Cuba, vol. ii., p 173.,
pl. xxiv., f. 1-3.

60. **Pl. solida** Adams, Contr. Conch. p. 61

MANGELIA Leach.

61. **M. quadrata** Reeve (Pleurotoma diminuta Adams).

Under this species as more or less distinguishable forms or varieties come the following: M. hopetonensis Reeve, plicata Adams, cervnia Adams and biconica Adams. Also Pleurotoma lavalleana and vespuciana Orb. Moll. Cuba.

DEFRANCIA Millet (Clathurella Carpenter).

62. **D. maculata** Adams.

Order FUSACEA.

MUREX Linné.

63. **M. pomiformis** Mart.

M. asperrimus Orb. (not Lam.) Moll. Cuba, vol. ii, p. 158.

64. **M. cornucervi** Mart.

M. brevifrons Lam. Orb. Moll. Cuba, vol. ii p. 159.

M. calcitrapa Lam.

M. purpuratus Reeve.

M. elongatus Lam.

M. florifer Reeve.

M. crassivaricosa Reeve.

M. toupiollei Bernardi.

M. microphyllus Lam.

65. **M. similis** Sow.

M. recurvirostris (Brod.) auctt.

M. messorius Sow. Orb. Moll. Cuba, vol. ii., p. 159.

M. tryoni Hidalgo.

M. rectirostris Sow.

M. beaui Petit, Journ. Conch. 1856, pl. viii, f. 1.

There are probably a number of other synonyms, but I would not venture on them without further information. The name *recurvirostris* belongs to the shell from the west coast of South America, which I hesitate to declare specifically identical with the West Indian one.

66. **M. nuceus** Morch.

67. **M. alveatus** Kien.

LAMPUSIA Schumacher (Triton auctt. not Linné).

68. **L. antillarum** Orb., Moll. Cuba, vol. ii., p. 161, pl. xxiii, f. 20.

L. tuberosum Reeve.

69. **L. succincta** Lam.

Triton americanum Orb., Moll. Cuba, vol. ii., p. 163, pl. xxiii, f. 22.

70. **L. martiniana** Orb., Moll. Cuba, vol. ii., p. 162. Murex pileare (Linn.) auctt.

RANELLA Lamarck.

71. **R. cubaniana** Orb., Moll. Cuba, vol. ii., p. 165, pl. xxiii., f. 24.

PERSONA Montfort.

72. **P. reticularis** Linn.

In this remarkable group (of which *P. anus* may be taken to be the type) there are several forms nearly allied, so that naturalists have often treated them as identical. *P. clathrata* is the eastern form while the West Indian one is that named by Linné *P. reticularis.* I cannot say if *P. cancellinus* Roissy is identical with *P. clathrata,* but it is not *P. reticularis.*

METULA Adams.

73. **M. lintea** Guppy, Proc. Scient. Assoc. Trinidad, December 1881, p. 178, pl. vii., f. 18.

This shell, allied to *M. cancellata* Gabb of the West Indian Miocene, was dredged in the Gulf of Paria by W. O. Crosby.

TURBINELLUS Lamarck.

74. **T. nassa** Gmel. Chenu. Conch vol. i, f. 913.

> T. cingulifera Lam. Kien. Coq. Viv. Tur-
> binellus 23.
>
> T. leucozonias Lam.
>
> T. carinifera Lam.
>
> T. dubia Petit, Journ. Conch. 1853, pl. ii, f.
> 9, 10.
>
> T. rudis Reeve
>
> T. brasilianus Orb., Voy. Amer. Merid. p. 449,
> pl. lxxvii, f. 17
>
> T. knorri Desh. Wood, Ind. Test. Murex 110.

75. **T. infundibulum** Gmel. Wood, Ind. Test
Murex 118.

> T. gibbulus Gmel.
>
> T. filosus Lam. Chenu, Conch. vol. i, f. 907.

FASCIOLARIA Lamarck.

76. **F. tulipa** Linn. Woodward, Man. Moll. pl. v, f. 1.

PYRULA Lamarck.

77. **P. melongena** Linn. Woodward, Man. Moll.
pl. v, f. 7.

78. **P. morio** Linn. Kien. Coq. Viv. Fusus 46.

> Fusus coronatus Lam.

The generic term *Ficula* should be reserved for the group of which *F. ficus* is the type. *Pyrula* is the proper name for the present subgenus of *Fusus* and *Rapana* is preferable for the group typified by *R. bulbosa*.

PUSIO Gray.

79. **P. articulatus** Lam. Chenu. Conch. vol. i, f. 617.

> Murex accinctus Born
>
> Buccinum pennatum Martini
>
>   ,,   plumatum Gmel.
>
> Purpura accincta Orb., Moll. Cuba, vol. ii, p. 146.
>
> Fusus articulatus Kien. Coq. Viv. Fusus 36.

TRITONIDEA Swainson.

80. **Tr. auriculata** Lam.

Buccinum coromandelianum Lam.

Pollia tincta Conrad

Triton caribeum Orb., Moll. Cuba, vol. ii, p. 162.

STROMBUS Linn.

81. **Str. gigas** Linn. Chenu. Conch. vol. i, f. 1583.

Str. goliath Chemn. Chenu, Conch. vol. i, f. 1581.

82. **Str. gallus** Linn. Chenu, Conch. vol. i, f. 1588.

83. **Str. pugilis** Linn. Woodward, Man. Moll. pl. iv, f. l.

Str. alatus Gmel. is very near to this and I think only a variety. The *Strombus proximus* of Sow. from the Miocene of Haiti &c. is also near but is distinguished (among other characters) by chevron-shaped color-bands, which persist in well-preserved specimens. *Strombus proximus* has denticulations inside the outer lip like *Str. bituberculatus.* (See a paper on the Miocene Fossils of Haiti in Quart. Journ. Geol. Soc. Nov. 1876. p. 521). *S. bifrons* Sow. from the same deposits is also near. *S. dubius* Sow. is the young.

84. **Str. bituberculatus** Lam.

Str. costosomuricatus Mart.

raninus Gmel.

affinis Gmel.

curruca Bolt.

lobatus Swains.

CASSIS Lamarck.

85. **C. testiculus** Linn. Chenu, Conch. vol. i. f. 1134

DOLIUM Lamarck.

86. **D. pennatum** Mart. (=D. perdix auctt.)

PURPURA Bruginères.

87. **P. patulum** Linn. Wood, Ind. Test. pl. xxii, f. 53.

88. **P. trapa** Bolt.=P. deltoidea Lam.

89. **P. bicostalis** Lam.

    P. hæmastoma Orb. Moll. Cuba, vol, ii, p. 144.

90. **P. undata** Lam.=P. bicarinata Kien.

    P. nodus Wood, Ind. Test. Murex 48.

91. **P. bispinosa** Phil. (Melongena). Journ. Conch. 1852, p. 157 pl. viii, f. 3.

    P. trinitatensis Guppy, Proc. S. A. Trin. Dec. 81 pl. vii, f. 17.

I thought of adopting the name of *P. floridana* for this shell and including the above names in the synonymy. But I do not feel quite satisfied on the point of specific identity, and the figures given by Tryon do not bear it out. The operculum is that of Purpura.

92. **P. gigantea** Reeve=P. consul Kien. (not Lam.)

    Buccinum hæmastoma Chemn.

93. **P. plicata** Mart.=P. galea Orb. (not Chemn.)

This is a Coralliophila and is the West Indian analogue of P. galea Chemn. The miocene relative is P. miocenica Guppy.

RICINULA Lamarck.

94. **R. turbinella** Kien, Woodward, Man. Moll. pl. iv, f. 15.

    R. nodulosa Adams=R. carbonaria Adams

    R. ferruginosa Reeve

95. **R. lugubris** Adams

PLANAXIS Lamarck.

96. **Pl. nucleus** Lam. Chenu, Conch. vol. v, f. 2143.

NASSA Lamarck.

97. **N. antillarum** Orb. Moll. Cuba, vol. ii, p. 141, pl. xxxiii, f. 1-3.

98. **N. (Phos) guadelupensis** Petit Journ.
    Conch. 1852 pl. ii, f. 3-4.

99. **N. (Phos) beaui** Petit Journ. Conch. 1856, pl.
    xii, f. 8-9.

TEREBRA Brugnières.

100. **T. flammea** Lam.

COLUMBELLA Lamarck.

101. **C. mercatoria** Linn. Woodward, Man. Moll.
     pl. vi, f. 10.

102. **C. laevigata** Lam.
     C. concinna Sow. Wood, Ind. Test. pl. xxi, f. 134.
     Voluta ocellata Gmel.

103. **C. argus** Orb. Moll. Cuba, vol. ii, p. 138, pl. xxi, f.
     34, 35.
     C. oscillatoria Sow.=C. cribraria Lam.

OLIVA Brugnières.

104. **O. reticularis** Lam.=O. olivaceus Mörch

105. **O. oryza** Lam.

106. **O. nitidula** Dillw.
     O. mutica Reeve.   O. monilifera Reeve

ETHALIA Adams.

107. **E. pusilla** Adams
     ? Odontostoma pusilla Pfeiff. (Zeitschrift 1851,
     page 88.)

I feel doubt as to whether this appellation is correct either generically or specifically. I only insert the name as a record of the existence in the Gulf of a shell having characters approaching to if not identical with those of *Ethalia*. I note a resemblance to *Rotella anomala* Orb., Moll. Cuba, p. 64, pl. xviii, f. 32, 33.

Order TROCHACEA.

ASTRALIUM Link

108. **Astralium rhodostomum** Lam.

TROCHUS Linné.

109. **Tr. byronianus** Gray

Tr. hotessicranus Orb., Moll. Cuba, vol. ii, p. 59,
pl. xviii, f. 15-17.

Tr. canaliculatus Orb. ibidem vol. ii, p. 60, pl.
xviii, f. 18, 19.

110. **Tr. excavatus** Lam.

Tr. umbilicaris Chemn.

ADEORBIS Wood. (Vitrinella Adams).

111. **Adeorbis striatus** Orb., Moll. Cuba, vol.
ii, p. 63, pl. xviii, f. 29-31.

Rotella semistriata Orb., ibidem, pl. xviii, f. 20-22.

„ diaphana Orb., „ „ f. 23-25.

„ carinata Orb., „ „ f. 26.28.

An extremely variable species. The above are a few names
given by D'Orbigny. How many more have to be added want
of reference to named specimens alone prevents me from saying.
Some of the forms recall *A. subcarinatus* Mont.

CYCLOSTREMA Marryat.

112. **C. schrammi** Fisch. Journ. de Conch. 1857,
p. 287 pl., x, f. 11.

PLEUROTOMARIA Defrance.

113. **Pl. adansoniana** Crosse & Fischer, Journ.
Conch. 1861, pl. v. Idem 1882, pl. i.

Though the single specimen I obtained in 1890 came
from Tobago I include it here as it probably inhabits the
channel between that island and Trinidad.[*]

---

* A full sized figure of this shell was published in an English
Conchological paper in 1892. An account of the shell was given in Proc.
Zool. Soc. Nov. 1891 page 484.

TURBO Linné.

114. **T. tuber** Linn. Chenu, Conch. vol. i, f. 2577.

NERITA Linné.

115. **N. peloronta** Linn. Wood, Ind. Test. pl. xxxvi, f. 46.
116. **N. antillarum** Gmel. Wood, Ind. Test. Nerita, 45.
   N. præcognita Adams.

NERITINA Lamarck.

117. **N. microstoma** Orb. Moll. Cuba, vol. ii, p. 47,
   pl. xvii, f. 35.
118. **N. meleagris** Linn.=N. chlorina Link.
   N. virginea Orb. Voy. Amer. Mer. pl. lvi, f. 1-3.
   N. jamaicensis Adams.
119. **N. viridis** Linn. Chenu, Conch. vol. i, f. 2460.

Order FISSURELLACEA.

FISSURELLA Lamarck.

120. **F. radiata** Lam. Wood, Ind. Test. Patella 101.
121. **F. alternata** Say.
122. **F. nodosa** Born. Chenu, Conch. vol. i, f. 2759.
123. **F. nimbosa** Linn, Chenu, Conch. vol. i, f. 2755.

EMARGINULA Lamarck.

124. **E. tensa** n. sp.

Subconic, moderately thick, white, with seven ribs running from the apex to distinct angles on the periphery of the shell, three of these ribs are much larger than the others and of these the central one covers the deep anterior sinus. There are two ribs of the second order and two of the third. Two other ribs run from the apex to the sides forming scarcely any angle. Between each of the nine principal ribs there are 5-8 finer costellæ. Apex smooth nearly central, slightly recurved backwards. Margins dentate, emarginate by a notch at the sinus. General contour roughly seven-angled and much wider behind than before. Interior lead-colored with a strongly marked sinus occupying the cavity of the largest rib.

More depressed than *octoradiata* and with much finer and more numerous less nodose ribs.

The lingual dentition indicates a near relationship to *Fissurella, Nerita* &c.

Order SOLENOCONCHIA.

DENTALIUM Linné.

125. **D. disparile** Orb. Moll. Cuba, vol. ii, p. 202, pl. xxv, f. 14–17.

CADULUS Philippi.

126. **C. obesus** Guppy.

CLASS CONCHIFERA.

Order PHOLADACEA.

PHOLAS Linné.

127. **Ph. candeana** Orb. Moll. Cuba. pl. xxv. f. 18–19.

128. **Ph. costata** Linn. Chenu, Conch. vol. ii, f. 1–3.

MARTESIA Leach.

129. **M. striata** Linn. Chenu, Conch. vol. ii, f. 48–50. Woodward, Man. Moll. pl. xxiii, f. 21.

TEREDO Linn.

130. **T. norvegica** Spengl. Woodward, Man. pl. xxiii, f. 26.

GASTROCHÆNA Lamarck.

131. **G. hians** Chemn. Wood, Ind. Test. Pholas 11.

G. cuneiformis Spengl. Chenu, Conch. vol. ii, f. 73.

Order ANATINACEA.

THRACIA Leach.

132. **Thr. dissimilis** Guppy, Proc. S. A. Trinidad, vol. i, p. 368.

Ann. and Mag. Nat. Hist. Jan. 1875, p. 52.

Thr. plicata Reeve (not Deshayes) C. J. Thraica 7,

PERIPLOMA Schumacher.

133. **P. inequivalvis** Schum.

134. **P. orbicularis** Guppy, Proc. Scient. Assoc.
Trinidad, vol. ii, (December 1881) p.
177, pl. vii, f. 13.

Order SOLENACEA.

SOLEN Linné.

135. **S. obliquus** Spengl. Wood, Appendix, pl. xi, f. 17.
S. ambiguus Lam. Chenu, Conch. vol. ii, f. 86.

136. **S. niveus** Hanley. Wood, Appendix, pl. xii, f. 40.

SOLECURTUS Blainville.

137. **S. caribeus** Lam.
S. gibbus Spengl. S. guinensis Dillwyn.

Order MYACEA.

CORBULA Brugnières.

138. **C. cubaniana** Orb. Chenu, Conch. vol. ii, f. 137.
C. knoxiana Adams, Contr. Conch. p. 238.

139. **C. caribea** Orb. Moll. Cuba, pl. xxvii, f. 5–8.
C. swiftiana Adams, Contr. Conch. p. 236.
C. kjœriana Adams, l. c. p. 237.

I feel sure that the synonymy of these species could be very
largely increased. The genus could be treated something after
the manner in which Krebs treated *Rissoina.* Most of Adams'
names are invalid.

NEARA Gray.

140. **N. ornatissima** Orb. (Sphena) Moll. Cuba,
pl. xxvii,f. 13–16.
Sphena alternata Orb. ibidem, f. 17–20.

Order MACTRACEA.

Mactra Linné.

141. **M. tumida** Chemn. Wood, Ind. Test. Mactra 8.
M. turgida Gmel. Chenu, Conch. vol. ii, f. 223–4.
M. guadelupensis Recluz, Jour. Conch. 1852, p.
249, pl. x, f. 4.

142. **M. pellucida** Chemn.

M. ovalina Lam. Wood, Appendix, pl. x, f. 23.

M. fragilis Wood, Ind. Test. Mactra 32.

M. brasiliana Lam. Wood, Appendix, pl. x, f. 60.

143. **M. alata** Spengl. Wood, Ind. Test. Suppl. Mactra 7.

M. carinata Lam. Chenu, Conch. vol. ii, f. 227–8.

Order LUCINACEA.

Iphigenia Schumacher.

144. **I. braziliensis** Lam. Woodward, Man. pl. xxi, f. 20.

DONAX Linné.

145. **D. denticulatus** Linn. Woodward, Man. pl. xxi., f. 19.

146. **D. striatus** Linn. Wood, Appendix, pl. xiv., f. 32.

These two Donaces appear to be almost always found together. Are they the two sexes of one species?

147. **D. pulchellus** Hanly Wood, Appendix, pl. xiv., f. 19.

D. powisianus Recluz.

ASAPHIS Modeer.

148. **A. deflorata** Linn.

A. rugosa (Lam.) Chenu, Conch. vol. ii., f. 256.

The *A. deflorata* of Linné is distinct from *A. rugosa* Lam. The latter inhabits the Pacific, while *A. deflorata* is the West Indian form. When conchologists find it such an easy matter to make a dozen species out of one it is regrettable that they cannot see the difference between two such species as these. But the fact is, mathematical and not natural characters are those which commend themselves as most susceptible of easy apprehension though of little value in nature. I may note that *Sanguinolaria taheitensis* (J. C. 1852, pl. x., f. 5) is a synonym of *A. rugosa* Lam.

SEMELE Schumacher.

149. **S. reticulata** Linn. Woodward, pl. xxi., f. 11.
150. **S. decussata** Gray. Wood. Ind. Test. Tellina 81.
    Amphidesma jayanum Adams.
151. **S. variegata** Lam. Wood, Ind. Test. Tellina, 17.
    S. purpurascens Gmel.

ERVILIA Turton.

152. **E. nitens** Turt. Woodward, Man. pl. xxi., f. 18.

TELLINA Linné.

153. **T. interrupta** Wood. Ind. Test. Tellina 4.
    T. listeri Bolten.
    T. maculosa Lam.
154. **T. punicea** Born. Wood, Ind. Test. Tellina 47.
    T. angulosa Gmel. Wood, Ind. Test. Tellina 64.
    T. striata Chemn. Wood, Appendix, pl. xiii., f. 10.
    T. alternata Say.
    Donax martinicensis Lam.
155. **T. rufescens** Chemn. Wood, Ind. Test. Tellina 37.
    T. operculata Gmel.
156. **T. constricta** Brug.
    T. interstriata Say.
    T. inornata Adams.
    T. gruneri Phil.
    T. cayennensis Lam.
157. **T. fausta** Dillw. Wood, Ind. Test. Tellina 74.
    T. remies Born (not Linn.)
    T. lawis Wood, Ind. Test. Tellina 68.
158. **T. radiata** Linn. Chenu, Conch. vol. ii., f. 272.
    T. unimaculata Lam. Wood, Ind. Test. Tellina 26.
159. **T. bimaculata** Linn. Wood, Ind. Test. Tellina 83.
    T. scaradiata Lam. Wood, Appendix, pl. xi., f. 37.
160. **T. exilis** Lam.
161. **T. candeana** Orb. Moll. Cuba, pl. xxv., f. 50-52.
162. **T. souleyetiana** Recluz, J. C. 1852, p. 253, pl. x., f. 5.

TELLIDORA Mörch.
163. **T. schrammi** Recluz, J. C. 1853, pl. vi., f. 7-8.

STRIGILLA Turton.
164. **Str. carnaria** Linn. Woodward, Man. pl. xxi., f. 6.

165. **Str. flexuosa** Say (Tellina mirabilis Phil.)

LUCINA Bruguières.

I have devoted some little attention to the genus Lucina and its highly involved synonymy and the following is the result of my researches so far as concerns the species found in the region to which this paper is devoted.

166. **L. pensylvanica** Linn. Woodward, Man. pl. xix., f. 6.

167. **L. jamaicensis** Spengl. Chenu, vol. ii. f. 566.

168. **L. pecten** Lam.
L. costata Orb. Moll. Cuba, pl. xxvii, f. 40-42.
L. textilis Phil.
L. antillarum Reeve.

169 **L. scabra** Lam. Chem, Conch. vol. ii. f. 576.
Tellina imbricata Chemn.
L. scobinata Recluz J. Conch. 1852, pl. x f. 6.
L. muricata Orb. (not Chemn.)

170. **L. muricata** Chemnitz.
L. scabra Reeve (not Lam.)

171. **L. nasuta** Conrad.
L. imbricatula Adams. Contr. Conch. p. 245.
L. obliqua Reeve (not Phil.)
L. pectinata Adams. Cortr. Conch. p. 245.
L. pecten Reeve. L. occidentalis Reeve.

172. **L. semireticulata** Orb. Voy. Amer. Mer. pl. lxxxiv, f. 7-9.
L. granulosa Adams. Proc. Bost. Soc. 1845.

173. **L. quadrisulcata** Orb. Moll. Cuba, pl. xxvii,
f. 34-36.

L. serrata Orb. Moll. Cuba, pl. xxvii, f. 37-39.

L. divaricata Lam. (not Linn.) Chenu, vol. li,
f. 572.

L. americana Adams. Contr. Conch. p. 243.

L. pilula Adams. Contr. Conch. p. 246.

L. chemnitzii Phil.

DIPLODONTA Brown 1833.

174. **D. brasiliensis** Mittre, Chenu, Conch. vol. ii,
f. 590.

175. **D. candeana** Orb. Moll. Cuba, pl. xxvii, f. 43-45.

LUCINOPSIS Forbes.

176. **L. tenuis** Recluz. Journ. de Conch. 1852, p. 250, pl.
x, f. 1.

DOSINIA Scopoli.

177. **D. philipii** Orb. Moll. Cuba, vol. ii. p. 270.

Cytherea concentrica Lam. (not Born)

„   patagonica Phil.

Order ASTARTACEA.

CRASSINELLA Guppy 1875.

178. **Cr. martinicensis** Orb. Moll. Cuba, pl. xxvii,
f. 21-23.

Crassinella guadelupensis Orb. l. c. f. 24-26.

In a paper by me in the publications of the United States
National Museum, I described a new species belonging to the
same genus as this under the name *Crassatella (Crassinella)
miocenica*. I inserted the following remarks under the description of the species. "As this paper does not deal with contro-
verted questions I will not go into that of the use of the name
Gouldia further than to remark that in addition to recent
specimens I have seen large numbers of the forms called *martinicensis* and *guadelupensis* by D'Orbigny from nearly every

miocene and pliocene deposit in the West Indies. The shell in both forms is particularly abundant in the pliocene deposit of Matura in Trinidad. My observations lead me to the belief that they are one species, and that they are adult, and not the immature form of any other species. And notwithstanding all that has been said I prefer to adhere to the name invented by myself *(Crassinella)* as the subgeneric name of the group, under *Crassatella."*

Order VENERACEA.

PETRICOLA Lamarck.

179. **P. typica** Jonas. Mollusk. Beitrag, pl. vii, f. 3.

180. **P. gracilis** Desh. Proc. Zool. Soc. 1853.

CYTHEREA Lamarck.

181. **C. hebraea** Lam. Wood, Appendix, pl. xiii, f. 21.
C. rubiginosa Phil.
C. varians Hanley. Wood, Appendix, pl. xv, f. 33.

182. **C. albida** Gmelin. Wood, Appendix, pl. xv, f. 31.
C. affinis Gmel.
C. læta Lam (not Linn.)

183. **C. circinata** Born. Wood, J. T. Venus 24.

184. **C. convexa** Say. Wood, Appendix, pl. x, f. 34.

185. **C. dione** Linn. Chenu. Conch. vol. ii, f. 378, 379.

Woodward's figure (Man. Conch. pl. **xx**, f. 8) represents the closely-allied but quite distinguishable *C. lupinaria* Lesson, which is an inhabitant of the West Coast of America.

TRIGONA Megerle.

186. **Tr. trigonella** Lam. Wood, Appendix, pl. xiii, f. 18.

187. **Tr. mactroides** Born. Wood, Ind. Test. Venus 33.
Cytherea corbicula Lam. Chenu. vol. ii, f. 385-7

Venus Linné.

188. **V. flexuosa** Linn. Chenu, Conch. vol. ii, f. 360.

V. macrodon Lam. Wood, Appendix, pl. ix, f. 7.

V. punctifera Gray.

V. auberiana Orb. Moll. Cuba, vol. ii, p. 277, pl. xxvi, f. 35.

189. **V. paphia** Linn. Woodward, Man. Moll. pl. xx, f. 6.

190. **V. cancellata** Linn. Wood, Ind. Test. Venus 6.

V. cingenda Dillwyn.

V. dysera Orb.

V. beaui Fischer, Journ. Conch. 1852, pl. xii, f. 15.

191. **V. granulata** Gmel. Wood, Ind. Test. Venus 29.

V. violacea Gmel.

V. marica Chemn.

192. **V. pectorina** Lam.

V. elegans Gray. Wood, Ind. Test. Suppl. pl. ii, f. 3.

193. **V. eximia** Phil.

?V. portesiana Orb. Voy. Amer. Merid. pl. lxxxiii, f. 1, 2.

194. **V. rugosa** Chemn. Wood, Ind. Test. Venus 13.

V. rigida Dillwyn.

Order CARDIACEA.

CARDIUM Linné.

195. **C. muricatum** Linn. Wood, Ind. Test. Cardium 14.

196. **C. leucostomum** Born.

C. elongatum Lam. Wood, Appendix, pl. xvii, f. 16.

C. subelongatum Sow.

C. marmoreum Lam.

197. **C. citrinum** Chemn. Wood, Ind. Test. Cardium 22.

C. serratum Linn.

C. lævigatum Lam.

198. **C. oviputamen** Reeve.

199. **C. eburniferum** Guppy. Proc. S. A. Trinidad, vol. i, p. 367. Ann. & Mag. Nat. Hist. Jan. 1875 pl. vii, f. 3.

This is near *C. isocardium* Linn. and may be a variety of that, but it has the vaulted scales much stouter and more crowded. It is common at Grenada.

CARDITA Bruguieres.

200. **C. pectunculus** Brug. Chenu, Conch. vol. ii, f. 655-656.

CHAMA Bruguieres.

201. **Ch. macrophylla** Chemn. Wood, Ind. Test. Chama 19.

Ch. imbricata Lam.
Ch. lazarus Lam. (not Linn.)
Ch. gryphoides Dillw. (not Linn.)

202. **Ch. florida** Lam.

203. **Ch. ruderalis** Lam.

Order MYTILACEA.

PINNA Linné.

204. **P. ramulosa** Reeve C. I. Pinna 52:

P. seminuda auctt. (as of Lam.)
P. rigida (Solander) Orb. Wood, Ind. Test. Pinna 7.
P. nobilis Chemn. (not Linn.)
P. orbignyi Hanley. Reeve, C. I. Pinna 49.

MYTILUS Linné.

205. **M. americanus** Favart.

M. modiolus Chemn. (not Linn.)
M. tulipa Lam. (part.)

206. **M. brasiliensis** Chemn. Wood, Appendix, p. 234.

     M. guainensis Lam.

     M. semifusca Sow.

207. **M. ligneus** Reeve.

This species spins a bag entirely covering the shell. It was found in 2-3 fathoms water on a bottom of sandy mud. Reeve gives no locality for his shell.

208. **M. exustus** Linn. Orb. Moll. Cuba, pl. xxviii, f. 6, 7.

     M. citrinus Chemn. Wood, Ind. Test. Arca 15.

     M. sulcata Lam.

209. **M. domingensis** Lam. Orb. Moll. Cuba, vol. ii, pl. xxviii, f. 8, 9.

     M. exustus. Lam.

210. **M. lavalleanus** Orb. Moll. Cuba, pl. xxviii, f. 3, 4.

     M. cubitus ? Say.

CRENELLA Brown.

211. **Cr. viator** Orb. Voy. Amer. Mer. pl. lxxxiv, f. 34-37.

LITHODOMUS Cuvier.

212. **L. bipenniferus** Guppy. Proc. S. A. Trinidad, Dec. 1877, p. 154. Idem Dec. 1881, p. 178, pl. vii, f. 14.

This species seems to have been confounded with its West African analogue, *L. caudigerus* Lam. with reference to which see Journ. Conch. 1865, p. 127. L. bipenniferus is more cylindrical; the umbones are usually less prominent and more nearly terminal; the dorsal edge has rarely so pronounced an angle. Burrows in limestone and corals.

AVICULA Bruguieres.

213. **Avicula vitrea** Reeve.

Specimens of this from deep water in the Caribean Sea were very kindly given me by Mr. W. H. Robinson and identified at the British Museum (Natural History). Since then some beautiful small specimens with finely devoloped lamellar spines have been found in the Bocas attached to the stem of a Gorgonid in company with that rare and extraordinary sea-star *Asteropora annulata*.

PERNA Bruguieres.

214. **P. obliqua** Lam.

> P. ephippium Sow. (not Linn.) Woodward, Man. pl. xvii. f. 2.
>
> Ostrea alata Gmel.

215. **P. chemnitziana** Orb.

> P. vulcella (b) Lam.
> Concha semiaurita Chemn.
> Perna bicolor Adams.

PECTEN (Gualtieri) Bruguieres.

216. **P. nucleus** Born. Wood, Ind. Test. Ostrea 47.

> P. turgidus Lam.

217. **P. nodosus** Linn. Chenu, Conch. vol. ii, f. 922.

218. **P. circularis ?** Sow. Thes. Conch. pl. xii, f. 23.

LIMA Bruguieres.

219. **L. scabra** Born. Chenu, Conch. vol. ii, f. 954-5.

> L. aspera Chemn.
> L. bullata Born.
> L. fragilis Lam.
> L. inflata Lam.
> L. glacialis Lam.
> L. pellucida Adams.

PLICATULA Lamarck.

220. **Pl. cristata** Lam. Woodward, Man. pl. xvi, f. 17.

Ostrea spondyloidea Meuschen.

Spondylus plicatus Chemn.

Pl. reniformis Lam.

221. **O. frons** Linn. Wood, Ind. Test. Mytilus 3.

O. folium Linn. (Reeve).

O. limacella Lam. Chenu, Conch. vol. ii, p. 1005.

O. rubella and erucella Lam.

222. **O. arborea** Chemn. Wood, Ind. Test. Ostrea 71.

O. radicum Chemn.

O. parasitica Gmel.

O. rhizophoræ Guild.

O. lingua and tulipa Lam.

223. **O. virginica** Gmel. Wood, J. T. Ostrea 68.

I insert this name on the strength of a specimen collected at Chacachacare.

Order ARCACEA.

LEDA Schumacher.

224. **L. jamaicensis** Orb. (var) Chenu, Conch. vol. ii,
　　　f. 901–903

225. **L. patagonica** Orb. Amer. Mer. p. 544, pl. lxxxii,
　　　f. 1–3.

226. **L. egregia** Guppy, Proc. S. A. Trinidad, 1881,
　　　p. 174. pl. vii, f. 1-2.

This resembles and may possibly be identical with *Leda new-combi* Angas from Colon. (P.Z.S. 1876 pl. xviii, f. 17.) It is a very fine species and of rare occurrence as it lives in 5-10 fathoms water.

227. **L. vitrea** Orb. Moll. Cuba, vol. ii, p. 262, pl. xxvi, f. 27-29.

    L. perlepida Guppy, Proc. S. A. Trinidad 1867, p. 173. Geol. Mag. 1874, pl. xviii, f. 9.

    L. vitrea, Proc. S. A. Trinidad, 1881, p. 172. pl., vii, f. 6.

NUCULA Lamarck.

228. **N. crosbyana** Guppy, Proc. S. A. Trinidad, 1881, p. 170, pl. vii, f. 3.

ARCA Linné.

229. **A. umbonata** Lam. Wood, Ind. Test. Arca 14.

    A. ventricosa Lam.

    A. mutabilis Reeve.

    A. americana Orb. (not Gray) Moll. Cuba, pl. xxvi, f. 1, 2.

    A. noæ auctt. (not Linn.)

230. **A. occidentalis** Phil.

    A. navicularis Brug. Wood, Ind. Test. Arca 5.

231. **A. listeri** Phil.

    A. fusca Brug.

    A. granulata Meusch.

232. **A. incongrua** Say. Wood, Appendix, pl. xviii. f. 44.

    A. brasiliana Lam. Wood, Ind. Test. Suppl. Arca 7.

    A. inequivalvis Brug.

233. **A. adamsi** Shuttl.

234. **A. americana** Gray. Wood, Ind. Test. Suppl. Arca 1.

    A. pexata Say. Wood, Appendix, pl. xviii, f. 43]

235. **A. centrota** Guppy, Proc. S. A. Trinidad, vol. i, p.
     368. Ann. & Mag. Nat. Hist. Jan. 1875, pl.
     vii, f. 4.

236. **A. floridana** Conr. (var.)

237. **A. auriculata** Lam. Chenu, Conch. vol. ii, f. 865,
     866.

238. **A. squamosa** Lam. (not Gray)

> A. donaciformis Reeve, Chenu, Conch. vol.
> ii, f. 863.
>
> A. divaricata Sow.
> A. domingensis Lam.
> A. angulata Meusch.

239. **A. reticulata** Chemn.

PECTUNCULUS Lamarck

240. **P. decussatus** Linn.

> P. pennaceus Lam.
> P. undatus Lam.
> P. angulatus Lam.
> P. hirtus Phil.

BRACHIOPODA.

DISCINA Lamarck.

241. **D. antillarum** Orb. Moll. Cuba, vol. ii, p. 368,
     pl. xxviii, f. 34-36.